WITHDRAWN

WITHDRAWN

THE MYTH OF HUMAN RACES

THE MYTH OF HUMAN RACES

ALAIN CORCOS

MICHIGAN STATE UNIVERSITY PRESS
EAST LANSING

Copyright © Alain Corcos

All Michigan State University Press books are produced on paper which meets the requirements of American National Standard of Information Sciences— Permanence of paper for printed materials
ANSI Z23.48—1984

Printed in the United States

Michigan State University Press
East Lansing, Michigan 48823-5202

05 04 03 02 01 00 99 98 97 1 2 3 4 5 6 7 8 9 10

Library of Congress Cataloging-in-Publication Data

Corcos, Alain F., 1925–
        The Myth of Human Races / Alain Corcos
            p.   cm.
Includes Bibliographical references and index.
ISBN 0-87013-439-6 (alk. paper)
1. Race I. Title
GN269.C69   1997
305.8—dc21   97-17769
                                                            CIP

THERE IS ALWAYS AN EASY SOLUTION TO
EVERY HUMAN PROBLEM,

NEAT, PLAUSIBLE, AND WRONG

H. L. MENCKEN

THIS BOOK IS DEDICATED TO

FLOYD V. MONAGHAN

A VERY FRIENDLY AND HIGHLY SKILLED
CRITIC

Contents

A Word to the Reader

You've got to be taught to hate and fear.
You've got to be taught from year to year.
It's got to be drummed in you dear little ear.
You've got to be carefully taught.
You've got to be taught before it's too late,
Before you are six or seven or eight,
To hate all the people your relatives hate.
You've got to be carefully taught.

Rodgers and Hammerstein
South Pacific

I am sitting in the living room of our cottage on the edge of Wiggins Lake, near Gladwin, Michigan. I am admiring one of those beautiful sunsets that generally soothe the mind, but in the background I am listening to a radio broadcast informing me that there is a resurgence of Nazism in Germany as skinheads are roaming the city streets and setting fire to houses of the people they hate, and that the German government is failing to nip this new Hitlerism in the bud, as it failed seventy years ago. As night falls, I become completely convinced of the necessity of writing this book which I hope will be a contribution to the fight against racism.

I have been fighting racism since, as a teenager, I was under the Nazi boot in Southern France during World War II. However, it is only in the last twenty years that, as a professor of biology at Michigan State University, I have had some influence. I taught a course dealing with the concept of race, stressing what biologists had to say about the subject. From the response of numerous students, I knew that I was reaching them. Many of them told me that, though the course had not changed their minds, for they were not racists before they enrolled in my class, they appreciated the course because it reinforced their beliefs and gave them ammunition with which to fight racism.

Since my retirement in 1991, I have thought of rewriting for the public at large the book I had written for my students in 1977. In this endeavor I was encouraged by my friends and colleagues. But after some thinking it became evident that a different work would be necessary, not so much because it would address another audience—after all, the public at large is not that much different from non-science-oriented students—but because during the past 15 years my thinking on that subject had evolved and I had come to the conclusion that there *were* no human races. The belief that there were created, on a scientific level, more confusion than clarity about human diversity. Of course I am not alone in having this view. It is widely shared today by many biologists and anthropologists, as well.[1] Now it is time for the public at large to abandon the belief in the existence of human races, a belief that has done great harm to human relations.

I realize, of course, that writing a book about the myth of human races will do little to change the true nature of racism. Racists will always find a good reason to hate people who do not look like them, who do not share their religious or political views, or who do not speak or dress the way they do. However, I believe that my speaking out is at least one step in the right direction.

Racism is based in great part on pseudoscience, half-truths, and untruths, and it is the scientist's task and duty to unmask and counter the irrationalities of racial prejudice with good science and honest interpretation. As a biologist, I am aware that it is difficult to dissociate one's responsibilities as a scientist from his or her duties as a private citizen, as well as from bias. I shall try to the best of my ability to present my arguments factually, the way I have learned about them. I believe that honest science and honest history are among freedom's most important weapons.[2] I am also convinced that the only way racism can be transcended is by educating people, especially children who "have to be taught to hate."

Notes for *Word to the Reader*

1. Alice Littlefield, Leonard Lieberman, and Larry T. Reynolds, "Redefining Race: The Potential Demise of a Concept in Physical Anthropology." *Current Anthropology* 23 (6): 641–647 (1982). Since then, most authors of new introductory books to Anthropology have abandoned the concept of race. Among them is Conrad Philip Kottak, *Anthropology: The Exploration of Human Diversity* (New York: McGraw Hill, 1991).

2. Ethan Alapenfelds wrote in *Sense and Nonsense About Race*: "Scientific facts alone will not make you love your fellow man, but facts will lay a firm foundation that can become the beginning of understanding. It is what you do after you know the facts that counts. Facts plus understanding plus a desire to conquer prejudice leads to constructive action."

INTRODUCTION

It is not the presence of objective differences between groups that create races, but the social recognition of such differences as socially significant or relevant.

Pierre L. Van den Berghe [1]

Historically, race thinking[2] assumes at least five things:

1. That humanity *can* be classified into groups using identifiable physical characteristics;
2. That these characteristics are transmitted "through the blood";
3. That they are inherited together;
4. That physical features are linked to behavior;
5. That these groups are by nature unequal and therefore can be ranked in order of intellectual, moral, and cultural superiority.

My objective in this book is to refute the notion that human "races" (in the plural) exist. I hope to contribute to efforts that will permit all of us to respond to human diversity in productive and useful ways.

Let us begin by taking a brief look at the first of the five assumptions mentioned above, which is the assumption that if human races did exist, we should be able to classify mankind into mutually exclusive groups. As we shall see in the following pages, no one has ever been able to succeed at this task because no one has ever found a specific set of characteristics that can be used to distinguish one group from another without introducing layers and layers of ambiguity.

The second assumption listed above, that blood is *the* vehicle of heredity, has been proven false. More than a century ago, research began to demonstrate that genes, not blood, transmit the hereditary information and that many genes express themselves differently in different environments. Yet, today people still talk in terms of "blood inheritance." For example, we often hear that a person acted in a particular way because of a hot blooded nature.

1

According to the blood theory of inheritance, each person is supposed to obtain half of his or her blood from each parent, one-quarter from each grandparent, and in decreasing fractions from remote ancestors. This view of inheritance led some who wished to promote racism to adopt the "one-drop" rule: Anyone who has a single drop of "black blood" is black; anyone who has a single drop of "Jewish blood" is a Jew.

According to the gene theory, on the other hand, each person is supposed to obtain half of his or her genes from each parent. However, it is impossible to predict what specific fraction of the hereditary contribution of each grandparent will be transmitted to each of their grandchildren. It could be as much as one-half, it could be nothing. At this time, it is impossible to define a parent's or grandparent's hereditary contribution, even after the birth of the child.

The third assumption is that groupings of physical characteristics such as blond hair, narrow noses, and blue eyes are inherited together. At first this appears to make sense. After all, we *seem* to have little difficulty in distinguishing Native Americans from dark-skinned sub-Saharan Africans, northern Europeans, or aboriginal people of Australia. What we are responding to is the fact that people from each of these areas *tend to have* some traits in common. But not all individuals in any given group have combinations of general traits in the proportions they should if this part of race theory were true. The physical traits being discussed are distributed throughout the entire human species; each trait is largely independent of others. They are not linked and therefore not inherited together.

The fourth assumption, that physical features are linked to behavior, is also without foundation. Two people differing in physical appearance will not necessarily differ in behavior. No valid, scientific proof has been offered to demonstrate that physical and mental characteristics are linked. Furthermore (and contrary to what has been shown to be true for ants, bees, or ducks) no one has been able to demonstrate that human behavioral characteristics are inherited. The belief that they are, and consequently that we cannot change them, has led to much pseudoscientific nonsense and to extreme horrors such as human slavery and the Holocaust.

The fifth assumption, that some human groups are by nature intellectually, morally, and culturally superior or inferior, has *never* been supported by scientific evidence, in spite of frequent and re-peated efforts of many who tried to find it.

It is sad that the idea of race is embedded in our historical con-sciousness and influences our political, economic, religious, recrea-tional and social institutions. Most of us have been conditioned during childhood to attribute a racial identity to ourselves and to others. With this in mind, words such as *white*, *black*, even *race* are used in this book as social terms. As biological terms, they are meaningless.

The belief in races, the race concept, may have had its origins in the events following European voyages of discovery to the New World and beyond. Popular ideas about differences between peo-ples grew out of European contact with the Indians of America, with Africans, and with peoples throughout Asia and the Pacific. Europeans were struck by the fact that the people they encoun-tered appeared to be physically different from themselves and they quickly began to devise systems of classification. These systems all emphasized differences between people based on skin color. Most Europeans also concluded that peoples of color were not as "civilized" as Europeans; therefore, they had to be inferior.

In other words, they were able to create a race concept and ra-cism that was based on the identification of combinations of physical differences, personal and social egotism, and political events. The conquest of indigenous peoples, their domination and exploitation, and the exportation of millions from Africa to the Americas to serve the insatiable greed of European entrepreneurs were their sources of justification. The physical differences, mainly skin color, were the major tools by which dominant whites con-structed and maintained social barriers and preserved economic inequality. Race thinking was born in Western Europe and ex-panded with the European conquest to infect countless minds, in-cluding those of racism's victims.

Europeans asked themselves what were the origins of the hereto-fore unknown peoples with whom they came into contact? They derived their initial answers from the Bible, which they believed contained clues to God's purpose for creating the world and every-thing in it. But the Bible said nothing about people inhabiting strange new lands, which Europeans at that time were still discov-ering. Why? Because the Bible itself had been interpreted by Euro-peans before their discovery of human diversity. Nevertheless, the Bible remained the framework for questions on the origin of humanity. For example, Europeans asked: Were newly discovered peoples descendants of Adam and Eve or did they represent another type of mankind? Their Biblically grounded inquiry led to many interpretations; most were to the detriment of non-European people.

In the seventeenth and eighteenth centuries a second wave of interpretation emerged, based on so-called scientific inquiry. In the same manner as they cataloged plants and animals, scientists tried to classify human beings. They looked for ways to categorize distinctive physical, biochemical, and intellectual differences. In spite of their efforts they were never able to find a difference. When one set of methods failed, they developed new ones. When faced with new scientific ideas, such as the evolution and gene theories, scientists seeking to establish a basis for proving the existence of more than one human race were compelled to adapt; but, once again, they failed.

Throughout this period of so-called scientific inquiry into the nature of human races, critics of the concept have not been heard. It has only been in the past twenty years that most scientists have at last rejected the idea of multiple races. They have realized that the reason they have not been able to classify humanity into multiple races is because *races do not exist*. Scientists have finally come to understand that the well-known factors that play important roles in race formation in plants and animals have not acted on human beings.

However, members of the public at large still believe in the existence of human races as firmly as they believe in the existence of planets. But there is no scientific connection. While planets are natural bodies that exist independently of humanity, races exist only in the minds of people. It is a sad fact that only recently, and I would say reluctantly, have scientists abandoned the race concept. Race is, and always has been, a social concept without biological foundation.

All of us, scientists included, cling to erroneous ideas; it is difficult to admit we have made a mistake. For a scientist, such an admission would force the abandonment of a particular line of research or force one to give up entirely on an idea.

But scientists can and do change their minds. One who did was astronomer Fred Hoyle. For thirty years he was a proponent of the idea that the universe is being continuously created. He finally embraced the opposite idea, that it had been created only once during what we now call the Big Bang. A more cynical reason for clinging to an erroneous theory was offered by Alexis de Tocqueville: "A false, but simple idea has more chance to succeed than a true, but complex one." ✓

In every field of science there have been erroneous ideas which have derailed scientific thinking for years. One of these was the phlogiston theory. Phlogiston was a substance invented by chemists to explain the diverse observations that accompany the burning of an object. This substance was supposed to migrate into and out of bodies undergoing a change into a different kind of material. The French chemist Antoine Lavoisier destroyed the concept of phlogiston by demonstrating that it was self-contradictory and unnecessary. The race concept has been said to be the phlogiston of anthropology.[3]

Scientists spent decades trying to establish an unambiguous race classification for human beings. In the process they demonstrated, often by accident, that the differences they tried to use to classify people were of no significance and that, whatever those differences might be, they are of no value. The differences we share are differences of degree and vary from individual to individual; they are not absolutes. This is true even of that most obvious "difference," skin color, which has played such a devastating role in shaping human relations. Scientists also have learned that there is more individual diversity, that is, diversity among individuals, than there is diversity among groups.

As we look at the world today, a world more and more crowded with people who may differ in appearance, in political, moral and religious beliefs, and in countless other ways, we will quickly realize problems we must face. If we wish to adjust our attitudes in this changing world, we must understand what the term "differences" means and what it does not mean. In particular we need to understand which things we are born with and which things are introduced by upbringing, education, and environment. These are the issues I address in this book.

Notes to the Introduction

1. Pierre L. Van den Berghe, *Race and Racism: A Comparative Perspective* (New York. John Wiley and Sons, 1967).

2 The phrase is that of Jacques Barzun, used in *Race: A Study in Superstition* (New York, Evanston, London: Harper and Row, 1965).

3. Ashley Montagu, *The Concept of Race*. (New York: The Free Press, 1964).

INTRODUCTION
TO PART ONE

You may think you are not qualified to decide whether or not many human races exist, or whether we are all part of a single race. You may believe that answering these questions should be left to scientists and others who have more information. Specialists have, after all, spent many years classifying humanity into racial groups; many have written scholarly articles and a seemingly endless stream of learned books on the subject. But notwithstanding all of what has been done, the means for understanding this seeming dilemma is within the reach of most people. We do not need to know about the structure of DNA, the detailed arguments for and against Charles Darwin's theory of evolution, or possess a scientific understanding of the intricacies of sexual reproduction. All we need to do is to open our eyes and look at people around us.

Can the people we see be classified conveniently into types and arranged by their skin color, from dark to light, as we have been led, too often, to believe? Or is human skin, regardless of color, all basically the same? Do people with light skin invariably have narrow noses and straight brown or blond hair, or do their noses and hair color vary from person to person regardless of skin color or lineage? Do people with dark skin always have brown eyes or does the color of their eyes vary as well? What about ear lobes, lips, hands, and other features? Observation will quickly lead us to conclude that human diversity is extreme and that those physical characteristics used historically to distinguish groups of people, Europeans, for example, are not unique to any single group.

Similar fallacies, such as the assumption that human racial characteristics are transmitted by blood, also must be challenged. We often hear stories, and are struck by the fact that characteristics identified with parents or grandparents can appear in children or grandchildren. A convenient, though false, explanation for this phenomenon is that the characteristics were transmitted by "blood." This observation (and many others like it) is also wrong. Transmission of human characteristics cannot be explained by a "blood theory of inheritance," according to which the bloods of the parents are mixed in the offspring.

Linking physical features to behavior and to race also must be contradicted. The entire range of human behavior is vast, and variations of behavior within groups, even in nuclear families (not to mention similar behaviors seen among people of different groups), belies any suggestion that links exist between physical appearance and particular kinds of behavior.

Finally, no evidence exists to suggest that human groups are by nature unequal or that they can be ranked in order of intellectual, moral or cultural superiority. In Part One, we shall demonstrate in detail that observation alone is sufficient to discard the fallacies outlined above.

Chapter 1

RACE IS A SLIPPERY WORD

The term "race" is one of the most frequently misused and misunderstood words in the American vernacular.

Peter I. Rose

Peter I. Rose[1] is indeed correct. Race, as applied to human beings, is vague and ambiguous. In common speech, it has a whole range of meanings. To focus on the issue, dictionaries offer little help. For example, the *American Heritage Dictionary of the English Language* gives numerous and contradictory definitions of the word "race":

1. A local geographic or global human population distinguished as a more or less distinct group by genetically transmitted physical characteristics.
2. Mankind as a whole [as in the human race].
3. Any group of people united or classified together on the basis of common history, nationality, or geographical distribution.
4. A genealogic line, lineage, family.
5. Any group of people more or less distinct from all others: the race of statesmen.
6. Biologically: (a) plant or animal population that differs from others of the same species in the frequency of hereditary traits: a subspecies; (b) or a breed or a strain of a domestic animal.
7. A distinguishing or characteristic quality, such as the flavor of wine.
8. Sprightliness, style.

These definitions are at variance with one another For instance, definitions 1 and 3 do not agree. One is a biological definition, the other a social definition. Definitions 2, 4, 5, 7 and 8 are taken from literary usage but have no validity in scientific or social science community. Definition 6 comes closest to the one used in this book. But, as we shall see later, much confusion about race has to do with how great the difference in the frequency of hereditary traits must be between two populations before we can label each of them a distinct race.

Race is a slippery word because it is a biological term, but we use it every day as a social term. In the mind of the public at large this leads to great confusion. Social, political, and religious views are added to what are seen as biological differences. All are seen as inheritable and unchangeable defining characteristics of one or another "race" of people.

However, most of the differences between groups are in fact cultural. People are of different national origin; they have different religions; they have different political views; and they speak different languages. These differences can be modified as shown by the fact that every day many of us change nationalities, religions, or political views. Most of us can learn language other than the one our parents taught us.

Race also has been equated with national origin. For example, writers and historians once spoke of the Roman or English races. However, no Roman race conquered much of the Mediterranean world, but rather people sharing Roman ideas of government, law, language, and military discipline. Among the Roman elites, there were many citizens who were not from the city of Rome or even from the Italian Peninsula, for that matter.

Similarly, there is a country called England, but there never was an English race. England was invaded by successive waves of peoples, many of whom differed from other invaders in physical appearance. Hence, no one can point to an English man or an English woman because, among the English, as among other Europeans, there are fair-skinned people, dark-skinned people, tall and short people, long-headed and round-headed people, people with long noses, people with broad noses.

Race also has been equated with religion. For instance, Jews are considered by many as belonging to a separate race. The history of the Jews is well known and reveals that originally they were nomadic people, a grouping of pastoral tribes the members of which spoke a Semitic language. They emigrated from the desert border of southern Mesopotamia to Palestine between the seventeenth and twelfth centuries BCE. On several occasions, they were expelled from the place they took as their homeland: a sojourn in Egypt terminated by the famous Exodus; the Babylonian captivity; the conquest by Rome. Thus, even before they were dispersed throughout the Roman Empire following the destruction of Jerusalem by Titus (70 AD), there were many occasions for breeding with other people of the Near East. Even if the Jews were originally a homogeneous group, which is unlikely, there has been extensive interbreeding with others from antiquity to the present day.

What Jews have preserved and transmitted is a rich body of religious and cultural traditions and modes of conduct. The only valid criterion for determining membership in the group is confessional (adherence to the Jewish faith). Jews are a religious body, not a separate biological human group.

Race has been equated with language, as in the case of "Aryan race." Historically, there was a common Indo-European language, called Aryan, from which Sanskrit, ancient Greek and Latin, and the majority of languages that are spoken today are derived. However, use of derivatives from that common root language does not mean that individuals speaking it look alike or hold similar religious beliefs. People do not speak Chinese, French, or English because of their biological inheritance but because their parents taught them the language, i.e., because of their cultural inheritance. In other words, whatever shape of mouth or vocal cords we have, we learn to speak the language that is spoken around us when we grow up. No matter what the color of our skin, shape of head, or the texture of our hair, we will acquire the language we hear in our homes or schools.

Though it is illogical to use religion, language, or nationality as a basis for inventing races, we continue to do so today. Many still regard Jews, Arabs, Mexicans, French, and Germans as being of different races. It may be more appropriate to consider some or all as ethnic groups, whose members share a common cultural background. Ethnicity is a social term. Race, on the other hand, is a biological term which can be used meaningfully only when applied to plants and animals, but never to human beings.

Even in biology, race is a slippery word. To understand why, we must understand how scientists use it and a related term, "species." Both race and species are used as categories for classification to reduce the vast array of diverse forms of animals and plants to manageable groupings for identification. However, there is a fundamental difference between what scientists mean when they identify a race and when they identify a species. Members of a species can breed with others of the same species but not with individuals belonging to different species. This gives the species a biological continuity and an exclusive membership. Species are real units in nature. Races, which are subdivisions of a species, differ from species in that their boundaries can never be fixed and definite because a member of one race can interbreed with members of another race.[2] It is, therefore, a capital error to believe that races have the same biological reality as species have. A species is a natural grouping of organisms, while a race is an artificial one whose definition is very vague and has changed with time without becoming clearer or more useful. This is especially true when applied to humans.

Race has been defined anthropologically as

". . . a great division of mankind, the members of which show similar or identical combinations of physical features which they owe to their common heredity." (Seltzer 1939)[3]

or genetically as

". . . a population which differs significantly from other human populations in regard to the frequency of one or more of the genes it possesses."[4].

The genetic definition seems at first more precise than the anthropological, because the frequency with which a gene occurs in a population can be calculated. However, it is important to realize that describing races on the basis of such frequency differences between populations presents a serious problem. For example, one wonders how great differences between two populations must be before they can be defined as distinct races. The frequency of a gene can run from 0 to 1, 0 meaning that a given gene is absent from a given population, 1 meaning that every member of the population has that particular gene. If the frequency of a particular gene was 1 in one population and 0 in another, we could argue that we have two distinct races. But in fact we have never been able to find such an example.

Both of the above definitions of a race, the anthropological and the genetic, are vague. They do not tell us how large divisions between populations must be in order to label them races, nor do they tell us how many there are. These things are, of course, all matters of choice for the classifier. It is no wonder that there was so much confusion among scientists when they attempted to classify humanity into distinct races.

Notes for Chapter One

1. Peter I. Rose, *They and We* (New York: Random House, 1964).

2. In a later chapter I will develop in detail the modern biological concepts of race and species. It will then become clear to the reader that according to such concepts there are no human races.

3. As quoted by Earl Count in *This is Race* (New York: Henry Schuman, 1950).

4. W. C. Boyd, *Genetics and the Races of Man* (Boston: Little, Brown and Company, 1950).

Chapter 2

RACE CLASSIFICATION: AN IMPOSSIBLE TASK

If an Arab and a Jew, standing completely naked in front of a large mirror, were to look at their bodies they would find that they were as alike as two peas in a pod.

Fernand Corcos (circa 1948)

What Fernand Corcos, my uncle, meant was that the two hypothetical men were more alike than they were different. As a matter of fact it would have been impossible for anyone, except the two of them, to know which one was a Jew and which one was an Arab; their "differences" would be purely cultural, not physical. My uncle could have made the same comment about the Serbs and the Croats: If they were fighting in the streets without their clothes on and without saying a word, they would be unable to recognize one another as "enemies."

Physical differences among other groups of human beings may be more varied than the differences between Jews and Arabs, or Serbs and Croats. For instance, we might be able to tell the difference between an Eskimo and a Watusi or a Bengali and a Swede; to deny these striking differences would be foolish.

But within an "obvious" difference there may lurk a subtle commonality. Good scientists know they should not always believe what they see; often, it is critical to go beyond direct observations, to look for interpretations which, though they may seem to deny the observations, not only explain them, but other things as well, things which at first appeared to have nothing in common.

Humanity is diverse; to understand this fact, all you need to do is spend a few minutes sitting in the terminal of an international airport. There we can observe people of different skin colors,

15

statures, hair forms and facial characteristics. Some people have dark brown skin, others have light skin; some have straight or wavy hair, others curly hair; some have prominent and thin noses, and others broad and flat noses; some have thin lips, others have large or everted lips. Some are tall and others short. Some have long heads, others round heads, and still others, intermediate heads. However, as we look longer and more carefully we begin to see that some individuals have more similarities with one another than they do with many others. In other words we can begin to see groups of individuals within the larger group. A sense of some kind of order begins to emerge from what we first saw as random and chaotic.

Since some of us are "more alike" than others, it seems only natural to attempt to place those who are similar into groups generally described as races. That is what scientists tried to do in the past. At first, they used indicators such as skin color, shape of eyes and noses, and so on. In the process, race became the framework for categorizing other ideas about human differences.

There is nothing inherently wrong with the idea of classifying. Grouping objects, events, ideas, or organisms is often helpful. It permits us to talk or to write about them in terms of their common traits, as well as their uniqueness. But for classification to be meaningful in the world of science, any member of one group has to be unequivocally distinguishable from a member of another group. We have no problem distinguishing a bird from a fish. Birds have feathers, most of them fly. No fish has feathers and none can fly. Hence, such classification is helpful. In the same vein, no mallard duck ever looks like a Canadian goose. Scientists are therefore justified in calling some birds mallard ducks and others Canadian geese.

A successful classification then provides us with an efficient tool to memorize the characteristics that an individual of a described group or class has. All we need to do is to remember the characteristics of the groups.[1]

At first glance, sorting humanity into more or less clear cut groups seems easy. If we think of a baker in France, a fisherman in Vietnam, and a peasant on the west coast of Africa, we have no problem imagining three men of different physical types whose ways of life are different, whose languages are not the same, and who follow different religions.

Similar distinctions were made in the past. Ancient Egyptians, for example, represented four types of people on the tombs of their Royal Dynasties: The first were black skinned with curly hair; a second type had a tinted creamy yellow skin and slanting eyes; another was light skinned, with blue eyes and blond beards; the fourth type, the Egyptians themselves, were people with reddish skin.

The division of mankind based on skin color is an old one. The first modern "scientific" classification schemes of this type did not appear, however, until the eighteenth century. At that time European thinkers were compelled to order nature into a mechanically logical system. Carl von Linné, the Swedish naturalist who originated the taxonomic classification system, divided humanity into four distinct groups: American or red, European or white, Asiatic or yellow, and African or black. The chief distinction made by Linné was skin color, but temperament, custom and habits were also taken into consideration. Here is his exact description of his four groups:

Americanus or red	Tenacious, contented, free, ruled by custom
Europeanus or white	Light, lively, inventive, ruled by rites
Asiaticus or yellow	Stern, haughty, stingy, ruled by opinion
Africanus or black	Cunning, slow, negligent, ruled by caprice

Linné,[2] in other words, linked physical traits, skin color, and behavior. This procedure is widely practiced today. When we see someone who is physically different from us, we expect that person to act and think entirely differently than we do. But, of course, such assumptions are wrong and no biological evidence exists that links behavioral and physical traits.

Georges Buffon, a French contemporary of Linné's increased the number of human groups from four to six. Buffon's groups included Laplanders, Tatars, Southern Asiatics, Europeans, Ethiopians and Americans. The German Johann Blumenbach, proposed to divide mankind into five groups: Caucasians[3], Mongolians, Malays, Ethiopians and Americans. Both Buffon and Blumenbach were fully aware of an element of artificiality that existed in their schemes for dividing mankind into groups, but most of their followers did not.

Throughout the nineteenth and into the early twentieth century, there were numerous attempts to classify mankind; all emphasized skin color, hair type, nose type and skull shape. Some classifiers recognized as few as three races, others more than thirty. These race classifications schemes all varied according to their originators' views, not only in regard to the group to which a particular population of human beings should be ascribed, but also in regard to the number of groups into which humanity should be divided.

The extreme difficulty of classifying mankind was stressed as early as 1787 by Samuel Stanhope Smith,[4] who wrote: "The conclusion to be drawn from all this variety of opinions is, perhaps, that it is impossible to draw the line precisely between the various races of man, or even to enumerate them with certainty and that it is itself a useless labor to attempt it."

Smith was absolutely right. Racial classification of human beings is impossible because humanity is diverse and distinct lines of demarcation between groups do not exist.

The difficulty in classifying mankind lies in the following: To serve as "racial" criteria, physical characteristics such as a particular pigmentation, hair texture, or eye shape must be present in the whole race in some fashion peculiar to that race alone. But no physical characteristic has ever been found to fit this criterion.

No matter how many groups we propose, we always find people who do not fit any one of the groups. Human diversity is so great that all attempts to divide mankind into "racial" groups have failed.[5] Why did earlier anthropologists get into such trouble when they tried to describe human races as separate and distinct groups? There are two reasons. First, they assumed that each race was created pure, i.e., that all members of any one race were all essentially alike in all characteristics. Second, they assumed that humankind as it exists today is the result of the mixing of races. Both assumptions are wrong.

It is a fact that pure races as envisioned by anthropologists earlier in the twentieth century do not exist now. For, wherever you go in the world, you find that only a very few, if any, inhabitants of any given region conform to anyone's proposed notions of racial purity. You will find, for example, that there are many tall, blond, blue-eyed persons in the city of Oslo in Norway, but you will also find in that city many individuals with brown eyes and dark hair. The opposite pattern of coloration is found in the city of Madrid, Spain, where blue-eyed and blond people are found among the brunette people who are generally considered as more typically "Spanish." One also can find blue-eyed and blond individuals among African tribes of the Congo and among Australian aboriginal groups. These individuals are not the result of having European ancestors, but are the result of changes that took place in their genes. Such changes can occur suddenly and infrequently and they are inherited. They are called mutations.

You could suppose that if there are no pure races now, there were pure races in the past. Such a supposition is a common one. But, likewise, it has no basis. The concept of pure race implies a homogeneous group of individuals. To obtain such homogeneity the group has to be small in number and it has to be isolated and allowed only to breed within itself for many generations. Through this type of breeding, called inbreeding, all the individuals become related sooner or later. The closest we have come to obtaining a homogeneous stock has been by mating male and female albino rats or mice for many generations. These rodents are used for cancer research where we want test animals as homogeneous as possible in order to obtain consistent results. These strains of rodents might be the only "pure races" in existence, but there are no pure races, even among domesticated animals. Even though breeding is controlled with the aim of producing homogeneous stocks, experience has shown that the resulting breeds may be fairly uniform for some selected features, such as gait or trotting in horses, coat color in dogs, egg-laying capacity in chickens, but these animals are highly diverse in other traits which were not selected for. If we cannot find pure races in domestic animals which are extensively inbred, one would expect them even less often in humanity where inbreeding has never been extensive (though it was more intense in the past than it is today). That pure races of human beings existed in the past is a myth.

Why have we believed in the existence of pure races for so long? Because such a notion reflects a way of thinking that we all possess. We have a tendency to think typologically, that is, we try to simplify things. If we have little experience with members of another group, we see very few differences among the individuals from within that group with whom we have contact. Hence, we generally conclude that "they all look alike." What we are doing is stereotyping. During the Korean war, a G. I., who is now a university professor of psychology, perceived that young Korean women were only of "one type" until one of them told him that she could not see any difference among the U.S. soldiers. To her they all looked like "G. I. Joes."

Stereotyping has done a lot of harm in human relations because it provides us with a warped picture of human diversity. Typological thinkers [6] ignore, or prefer to ignore, variability among individuals and imagine that an ideal, or average type exists for each real or imagined human group. To them, the average types are "real," not the individuals. For example, a educator thinking typologically thinks of an average student. But, of course, there is no such student because every student is different. We hear often of the typical scientist, the typical boss, the typical Southerner, the typical voter. All are products of typological thinking. They have two things in common: They stand out as a sort of example of all the individuals of a group and they do not exist in reality.

Anthropologists of the eighteenth, nineteenth and early twentieth century attempted to classify mankind, but failed, because— and it is worth repeating—fundamentally our diversity is greater among ourselves as individuals than among groups into which we happen to be classified. To be successful, a classification scheme presupposes that some objects that are judged have more similarities among them than with others. Otherwise, there would be no way to place them in separate groups. When anthropologists began to classify mankind into races, they used as their starting point the observation that some of us looked more alike than others; they also assumed that these similarities could permit classification without ambiguity. If their starting point were true, their assumption was not. In science, assumptions are useful if they lead to the understanding of a natural process. But when there is no evidence to support an assumption or evidence exists to contradict it, then the assumption should be abandoned.

This is what anthropologists of the nineteenth and early twentieth centuries should have done in the face of evidence that it was impossible for them to agree on a racial classification system. Unfortunately, they did not question the fundamental assumptions on which their classification theories were based and they never abandoned their belief in the existence of human races.[7] A lonely voice, however, may have been heard as early as 1906. Jean Finot[8] warned that "races are irreducible categories existing only as fictions in our brains."

Notes to Chapter Two

1. For an excellent analysis of the use of categorization, see *A Study of Thinking* by Jerome S. Bruner, Jacqueline J. Goodnow and George A. Austin (New York: John Wiley and Sons, 1956).

2. Linné at one time included in his famous book, *Systema Naturae*, another species besides ours, that he called *Homo monstrosus*. He seemed to have shared the belief with so many other people that there were semihuman creatures living in remote parts of the world: some were headless, some had one eye, some had only one foot which they used as an umbrella, some had two goat feet, some had big lips or long ears.

3. The term Caucasian, still used by the police to describe the appearance of alleged criminals, can be equated with the term "white." It is part of an old and antiquated classification system, which we are discussing in the following pages.

4. Smith, Samuel Stanhope, *An Essay on the Causes of the Variety of Complexion and Figure in Human Species.* (Cambridge: Harvard University Press, 1965). Originally published in 1810.

5. For example, are the Turks Europeans, as their appearance suggests, or do they belong with the Asian tribes of Central Asia, with whom they have linguistic affinity? What is to be done with the Basque. who appear to all eyes to be Spaniards and yet whose language and culture are seemingly unrelated to any other in the world?

6. When he was governor of California, Ronald Reagan is reported to have said: "If you have seen one redwood, you have seen them all." This statement reflects the mind of a typological thinker, but it also could reflect the fact that Reagan just did not care about redwoods.

7. Though classification of mankind is impossible, it is necessary to refer to the various human populations that exist throughout the world. In this book we have given names to these populations. Those names are used as descriptive tools, not as part of a classifying scheme. 1.) Europeans: populations of Europe, North Africa, and the Middle East. 2.) East Indians: populations of the Indian subcontinent. 3.) Asians: populations of Siberia, Mongolia, China, Japan, Southeast Asia and Indonesia. 4.) American Indians: the descendants of the native populations of America before the coming of the Europeans. 5.) Africans: populations of Africa. 6.) Australian aborigines: the descendants of the native populations of Australia before the coming of the Europeans.

8. Jean Finot, *Race and Prejudice*. Trans. by Florence Wade-Evans. (London: Constable and Co., 1906).

Chapter 3

SKULLS, WOMEN, AND SAVAGES: THE ART OF CRANIOLOGY, OR HEADS, I'M SUPERIOR; TAILS, YOU'RE INFERIOR[1]

The belief that there exists in man some close relationship between the size of the brain and the development of the intellectual faculties is supported by the comparison of the skulls of savages and civilized races, of ancient and modern people, and by analogy of the whole vertebrate series.

Charles Darwin[2]

Charles Darwin, though highly intelligent and an original thinker, shared a belief that brain size was highly correlated with intelligence; that is, the larger the skull and the bigger the brain, the more intelligent the person. Darwin and his colleagues also firmly believed in the existence of human races. Many thought that these races, like different species, had been created separately and endowed with different mental capacities.[3] Others thought that human races had evolved from a common ancestor, but that, by now, had acquired distinct physical and mental capacities. Whatever they thought about the origin of human races, most intellectuals of nineteenth century, scientists and philosophers alike, were convinced of the superiority of the white race. This conclusion was not surprising considering the fact that these intellectuals were either Europeans or of European ancestry. Convinced of their superiority, they chose to study two characteristics they believed were the most likely to be related to intellectual development: skull shape and brain size.

23

Where did they get the idea that brain size was correlated with intelligence? Very likely, from the fact that human beings have one of the largest brains of any species that exists on the earth today, or that may have existed in the past. Only elephants and whales have brains that are larger than our own. On the other hand, our brains are about two to three times the size of those of the big apes. Since we are able to do far more things than they do, it was natural to think that there might be a relationship between the size of brain and the mental abilities among different animals.

The contemporaries of Darwin extended this idea of relationship of brain size and intelligence to human beings. Convinced that there was such a correlation, they decided that the best scientific procedure to support their idea was to measure skulls of living people and to weigh brains of the dead. They even created a new science, craniology; it soon became a hot subject, not only in technical journals but also in the popular press. The tools of this new science were calipers and lead shot, the latter for pouring into empty skulls for determining capacity.[4]

It is not hard to see why scientists picked the skull as the object of study. The skull is exceedingly durable. It is found in a fair state of preservation when the rest of the body, including most of the skeleton has disappeared. Also, the skull, by its conformation, offers opportunities for making numerous measurements with a fair degree of precision.

At first there seemed to be some evidence that these ideas might be correct, for certain populations tended to be more broad-headed than others and this encouraged anthropologists to believe that racial groupings could be described on this basis. The relation between the breadth and the length of the skull was considered for a while to be the most important criterion for racial distinction. This relation was expressed as the cranial index which is derived by dividing the width of the head by the length of the head. If, for instance, a head was 150 millimeters wide and 200 millimeters long, its cranial index was 0.75. Any long and narrow skull whose cranial index was below 0.75 was called dolicocephalic. Any short and broad skull whose cranial index was above 0.80 was called brachycephalic. Skulls from between these limits were called mesocephalic. Craniologists went so far as to make believe that longheads and shortheads were of two different races, with the mesocephalics being hybrids between the two. Soon the anthropologists, convinced that rigorous measurements were the secret of precise science,[5] had craniology resting on a mathematical framework and every human group was catalogued as to their average cranial index.

Anthropologists collected skulls from many parts of the world. The most famous collection was made by Samuel George Morton, a prominent nineteenth century Philadelphia physician. He measured with great accuracy hundreds of skulls that he had classified into the following groups: Caucasian, Mongolian, Malaysian, American and Ethiopian. He summarized his data in a table which indicated that the average cranial capacity of each of these groups was different, the greatest being that of the Caucasian and the smallest that of the Ethiopian.

For years no one disputed Morton's conclusions. Few of his "white" scientific colleagues doubted that "whites" were superior to Indians and blacks.[6] However, today one cannot take Morton's results seriously for it is evident that he was convinced that the "white race" was superior to any other. He also finagled the statistical data by convenient omissions, miscalculations, shifting criteria, separating male from female skulls in some samples, but not in others. He seems to have done this unconsciously, for he published all his raw data, which when reinterpreted, as Gould did in 1976,[7] showed that all "racial groups" had approximately equal cranial capacities.

There was no need to wait for Gould's reinterpretation of Morton's data to abandon the idea that there was a correlation between skull size and brain capability. Such an idea was an untenable one, easily contradicted by the facts, even at the time of Morton. Not all great scientists, artists, and writers have large heads or big brains. For example, there was a tremendous range in brain weight among the well-known intellectuals of the nineteenth century, from just above 1000 to more than 2000 grams (28.3 grams equal an ounce). Some notable deviations from the average weight of human brains (1350 grams) were (in grams): Thackeray, the novelist, 1644; Cuvier, the comparative anatomist, 1830; Turgenieff, the historian, 2012; Haeckel, the biologist, 1575; Agassiz, the zoologist, 1495; Schumann, the musician, 1475; Gambetta, the statesman, 1294; Whitman, the poet, 1283; Dollinger, the anatomist, 1207; Wagner, the musician, 1855; Gauss, the mathematician, 1492; and Anatole France, the writer, 1017.

Obviously, it is not the size or the weight of a brain that is important, but how it functions. There were political problems as well as biological problems with craniology. For instance, Paul Broca, professor of clinical surgery, who founded the Anthropological Society of France, firmly believed that there was a nice relationship between the size of a brain and its intelligence. However, he found himself one day in a serious predicament. One German anthropologist, named Ernst Huschke, claimed that the weight of German brains was, on average, one hundred grams heavier than French brains. Broca came to the rescue of French patriotism. He used correction factors to decrease the differences between the German and the French brains. One of them was valid, namely that brain size increased with body size. Since Germans were, on average, taller than Frenchmen, it was natural for Broca to correct for size. Once he had done this, the average German brain was smaller than the French. The honor of France was saved.

To believe that there were German and French races, as Broca and Huschke did, was as illogical as to propose that people of California and those of Ohio were two different races because their cephalic indexes differ by a few points. One of the first serious blows struck to craniology came from Karl Pearson, a world-famous student of biological measurements. He demonstrated that there was no correlation between skull capacity and intellectual power. Among the examples he used was a comparison between cranial measurements of Cambridge University undergraduates and their examination results. The correlation was nonexistent.

However, the reason why craniology finally collapsed had a lot to do with the discovery that the shape of the skull is very much subject to environmental factors, as shown by the fact that populations of Europe and Asia are slowly changing from being long-headed to round-headed. So, if indeed, as it was once postulated, longheads were more intelligent and industrious than roundheads, the European population is in a rather bad way. Europeans must be getting more and more stupid all the time.

What environmental factors have changed to cause the shape of the European skull to change from dolicocephalic to brachycephalic, we do not know for sure. What we do know today is that bone is a very plastic material during the period of growth. Its form seems to be readily modified by dietary deficiencies or other environmental disturbances of one sort or another. This might explain the discrepancies the craniologists found in their data. But what is more important, it destroyed one of their fundamental assumptions.

If shape of the skull was not the result only of heredity but was influenced by the environment, it could not be a characteristic of a race.

Craniologists also tried to use their science to demonstrate that not only were northwestern Europeans intellectually superior to other human groups, but that men were superior to women.[8]

The movement reached a high point in 1868, in England, the first country to see the beginnings of an active women's movement. Women claimed the right to vote; they demanded education and clamored for admission into the universities. But there were men who were far from ready to grant these rights. They argued that the concept of intellectual equality of men and women was absurd. Members of the London Anthropological Society, all males, were sure that the subordinate position of the women was based on mental inferiority. Now was the time, they said, to demonstrate scientifically that their position was correct. They would make precise comparative measurements of the skulls of the sexes.

These scientists were more biased, if that is possible, about the intellectual abilities of women than they were about those of non-whites males. That they had made up their minds before they began their research is demonstrated in the following statement made in 1868 by Luke Owen Pike [9]:

> Must not woman's intellectual capacity be as different as her anatomy? If there is truth in science, the intelligence of a woman not only has but must have a certain relation to her structure and if it could be shown that there exists no difference between male and female minds, there would not be anthropology.

A year later, another anthropologist, McGrigor Allan,[10] said that women's brains were analogous to those of animals:

> In them the organs of sense were overdeveloped to the detriment of the brain proper. This explained the observed fact that women were sensuous and emotional, less guided by reason than men. If women were to be educated, their innate intuitive faculties might be destroyed.

Female skulls were submitted to the craniological calipers. At first the decision to base the anatomical argument for female inferiority on cranial size and shape seemed promising. On the average, female brains were found to weigh less than male brains and female crania, smaller than male crania. No surprise here. If intelligence depended on cranial size, women were less intelligent than men. However, the logic of this argument had one fatal flaw, which is known as the "elephant problem." If either the absolute size of the brain or cranial volume was to be taken as a measure of intelligence, then the elephant and the whale must be the lords of creation because they possessed brains much larger than ours. Some craniologists tried to finesse the elephant problem by asserting that the relative weight of the brain to the body was the true measure of intelligence. We were smarter. Our honor, as a species, was saved.

When this method was applied to women, who, in general, have a skeleton smaller than their male counterparts, it was discovered that in relation to body weight the female brain was larger than the male's. This is also the case with children, birds and small monkeys. However, instead of accepting the measurements as they were, anthropologists continued their attempts to prove women inferior by taking other types of measurements. But these also failed to support their beliefs.

English people were not the only ones who believed women to be inferior; the French did, as well. Paul Broca believed women to be inferior and based his belief on the smaller size of the woman's brain. He had collected many brains, male and female. The mean weight of his male brains was 1325 grams and that for female brains was 1144 grams, a difference of 181 grams, or 14 percent. Let us remind the reader that Broca was the same man who had rescued the honor of France by using a factor that corrected for the larger size of the German's. He seemed to have known, therefore, that part of the difference between male and female brains had to be attributed to the larger size of the males. But strangely, Broca did not make any attempt to correct for size in his female data and actually stated that he did not need to do so. He simply stated that size cannot account for the entire difference because we know that women are not as intelligent as men. Anthropologists of the nineteenth century were so sure that women and "nonwhites" were inferior that, when cranial index or brain weight studies indicated that they might be wrong, they simply looked for other types of measurements to support their preconceived notions.[11]

For a while, cranial height was used as a criterion for intelligence. But this criterion was soon disputed by John Cleland, who warned that since Kaffirs, blacks, and Australian aborigines—all considered inferior peoples—possessed high foreheads, cranial height could not be regarded as a criterion of superior intellect. After having stated that, Cleland looked for another index to measure intelligence: he chose the angle between the baseline of the skull and the vertebral column, but he did not get any place with that index either. In this respect, there were no differences between men and women. If there were differences in intelligence between men and women, they could not be demonstrated by craniology.

With time it became obvious that there were no skull differences between "whites" and "non-whites." The basis for craniology crumbled and this pseudoscience collapsed, and with it another method of classifying human beings into "racial" groups.

But the urge to prove that there are differences in intelligence between sexes, human groups, and classes is not dead. Even today, there are people who want to prove by any means that women or "blacks" are inferior and less intelligent. They have not given up. They have simply turned to other methods and a more sophisticated form of craniology. One of these methods is the IQ test, whose nature we shall discuss in a later chapter. Craniology and intelligence testing of human groups fulfill similar functions in reinforcing and rationalizing existing social divisions. As long as there are entrenched social and political distinctions between sexes, human groups, and social classes, there will be forms of science or rather pseudoscience,[12] whose main function is to rationalize and legitimize these distinctions.

The pattern that we have seen here of an initially plausible assumption being destroyed by further accumulation of data is one we shall see repeatedly in the following chapters.

Notes

1. After S. J. Gould, *The Mismeasure of Man*. (New York: W. W. Norton and Co., 1981)

2. In Charles Darwin's *The Descent of Man*, 1971. In dividing skulls into savage and civilized, Darwin had already decided how intelligence should be gauged. In other words, he already assumed that civilization produced larger skull size. He omitted the consideration that the development of intellectual faculties should ideally be measured within the context of one's culture, i.e., the savage may be better adapted intellectually to his physical environment.

3. This was true for Louis Agassiz, one of the most famous biologists of the nineteenth century.

4. At first white mustard seed was poured into the skull cavity, but it was hard to pack the seed, which led to imperfect measurements. Lead shot of one-eighth inch diameter was found to be more practical and this became the standard way of measuring brain size.

5. Science is more than taking measurements or making calculations. What is far more important is what the scientist does with the facts he or she has at hand: drawing conclusions, making generalizations, and testing predictions (hypotheses).

6. Myths die hard. In the eighteenth edition (1964), of the *Encyclopedia Americana*, one can still read that a small brain in relation to a person's size is a characteristic of "black" people.

7. Stephen Jay Gould, "Morton's Ranking of Races by Cranial Capacity: Unconscious Manipulation of Data May be a Scientific Norm," *Science* 200 (5 May, 1978): 503-509..

8. I am deeply indebted to the following article by Elizabeth Fee: "Nineteenth-Century Craniology: the Study of the Female Skull." *Bulletin of the History of Medicine* 53(1979):415-433..

9. Luke Owen Pike, "On the Claim of Women to Political Power," *Journal of the Anthropological Society* 7(1869): iv.

10 Allan McGregor, "On the Real Differences in the Minds of Men and Women." *Journal of the Anthropological Society* 7(1869): cxcvii.

11. A Philadelphia lawyer, Peter A. Browne, became interested in the structure of animal and human hair. Armed with a microscope and a little instrument he invented called a trichometer, he was able to discover that all human hair was not alike. Some was oval, some cylindrical, some elliptical. He believed he could associate oval hair with Europeans, cylindrical hair with Indians, and elliptical with Africans. There were other differences which convinced him that there were real differences among human "races" and that these "races" had separate origins. Though his hair collection was large and his research was meticulous, he did not have much influence in anthropology.

12. The type of question a scientist asks is, whether something is true. The type of question a pseudoscientist asks is, how can we prove this?

Chapter 4

FULL BLOOD, HALF-BLOOD, AND TAINTED BLOOD

"It ain't necessary so."

Porgy and Bess

I am spending a weekend on Wiggins Lake, in northern Michigan. It is cold, but I feel fortunate that I am not in the eastern United States, which is being battered by what the press calls the worst storm of the century. Inside our cottage I am warm and watching a "thrilling" movie called *Tainted Blood*. But as the film's plot unravels I become disenchanted with it, because I realize that the movie would be thrilling if I did not know that the three common beliefs about heredity on which it is based are without foundation. What is wrong with *Tainted Blood* is that there is no such thing. Heredity is not transmitted by blood, and there is nothing inherently special about a boy and a girl just because they are born at the same time. Although they are twins, they are not identical and therefore if one has a murderous disposition, there is no reason to expect the other to behave the same way. These common beliefs are among the many myths that science has rejected but which still persist in the minds of the public at large and obviously also in the minds of some mystery writers.

Even today, people know little or nothing about the mechanisms of inheritance.[1] As a result, they often make vital decisions based on ignorance. Here is but one example: years ago, I lived in a small town in Oregon. My next-door neighbors were friendly, and during one of our initial conversations, I asked if they had children. They said no. The husband added: "Three years ago my wife learned that she was descended from black slaves on her father's side. Her fear of having a Negro baby has kept us from having children." Here was a young couple missing the joys of parenthood because they believed a myth about the appearance of what are commonly called "throwbacks"; in this case, they believed it was possible for them to have a black child. Being a geneticist, I tried to explain to them as best I could that such a thing could never happen. I never learned whether or not I convinced them or if they had children, for I left the town ten months later. Yet, I wonder how many couples' lives have been, and still are, affected by this and the many other myths that surround heredity.

This personal encounter with what people believe to be true made me realize that, if we are ignorant of the true nature of heredity, we cannot understand the true nature of the problem of race. Some people persist in believing not only that the characteristics of the child are the result of the blending of the parents blood at conception,[2] but also that someone's "race" is inherited, when in fact *genes* are inherited; genes that dictate skin color, hair, the shape of nose or lips, and so on.

According to ancient and fallacious theories of inheritance, the physical means by which transmission occurred were thought to be the commingling and blending of parental substances in the offspring. Something of each parent, popularly referred to as "blood," was assumed to lose its own individuality in the blend, which occurred in the child, and this blending process repeated itself in the children's children and in later descendants. Each person was supposed to have obtained half of his or her inheritance from each parent, one-quarter from each grandparent, and so on in decreasing fractions from remote ancestors. This repeated halving is represented in the following table.[3]

Table 1. An example of terminology for members of "mixed races" and their constitution in terms of "blood" fractions

Parents	Offspring	Degree of mixture
Negro and European	*Mulatto*	*1/2 European 1/2 Negro*
European and Mulatto	*Terceron*	*3/4 European 1/4 Negro*
European and Terceron	*Quarteron*	*7/8 European 1/8 Negro*
European and Quarteron	*Qinteron*	*15/16 European 1/16 Negro*

It is interesting to ask why blood was ever thought to be the vehicle of heredity. Perhaps the idea was closely related to the fact that the fluid is absolutely necessary for survival. Since earliest times observers have known that human beings become weaker and sometimes die after appreciable losses of blood. This fact led our ancestors to regard blood as the essence of life itself, endowed with strength-conferring qualities as well as magical properties. For example, in order to make seeds fertile, some groups sprinkled them with blood before planting; in order to transmit bravery, wisdom, or other qualities from their enemies, some groups drank their blood; in order to establish unbreakable bonds of friendship, some exchanged blood or drank blood from a common source and in this way became "blood brothers."

Assuming blood to be the carrier of heredity, where was the mixing of the male blood and female blood supposed to take place? The Greek philosopher Aristotle had an idea.[4] He proposed that male semen was produced from the blood, that it was in fact highly purified blood. The menstrual fluid of a woman was her semen, but it was not so highly purified as that of the male because women were weaker, "colder" than men, and thus they did not have the power to cook their product completely. When the two semens united, the "less concocted" semen of the woman furnished the substance of the embryo, while the male semen added the form-giving property. In other words, the female provided the building material, while the male provided the life-giving power that formed its material into an embryo and gave it life.

Aristotle compared the male contribution to the work of a carpenter and the female contribution with the timbers of which things were made. Readers may object that Aristotle's view of the role of women in the creation of children carried a male bias. Whether or not that is true, his explanation also had a serious defect. If the link between semen and concocted blood explained the overall resemblance between parents and offspring, it made no provision for the detailed cases of resemblance such as when a girl has her father's crooked nose or when a boy has his mother's allergies. Yet it is precisely these kinds of cases that have to be explained in any theory of heredity. Aristotle did not give us any mechanism by which they were possible.

As a matter of fact, the blood theory of inheritance fails to explain many things. For example, it does not explain the extreme diversity of humanity that occurs generation after generation. Such diversity is contrary to what the blood theory predicts, that is, that with each passing generation, the descendants of parents should become more and more alike; but they do not.

Let us take skin color as an example. If the blood theory were correct, all shades of skin color would quickly disappear and human beings would be of only one medium shade. To use an analogy, let us take four ounces of red-colored water and four ounces of clear water and mix them. The result will be eight ounces of pink water. If we add another eight ounces of pink water, we will have sixteen ounces of pink water. If we were pouring the mixture into eight two-ounce glasses, all the glasses would contain pink water—not some glasses getting pure red water, some getting pink water, and some getting colorless water. In other words, the mixture will not spontaneously revert to pure red or pure white. The same is true of two different batches of paint; once mixed there is no way of unmixing them. The blood theory of inheritance may predict correctly the results of mating between a very dark-skinned individual and a very light skinned one. However, it fails to predict the skin color of the children of two medium-skinned parents, which may range from very dark to very light.

The theory of blood inheritance also does not explain why a child often has characteristics which neither parent possesses, but which one of the grandparents, or even a more remote ancestor, may have had. For example, some children are born albino whose parents are normally pigmented, while one of their grandparents was albino. It appears that albinism has skipped a generation. The

blood theory of inheritance could not explain this fact nor could it explain why brothers and sisters, though showing family resemblances, are clearly different from one another. One may be tall, brown-eyed and have normal vision while another may be short, blue-eyed and colorblind. Though many biologists never accepted the blood theory of inheritance because of its failure to predict correctly or to explain observations, it took a long time for the theory to be entirely discarded.

Today, however, we know that blood is in no way connected with the transmission of hereditary characteristics. This task is the function of genes which lie in the chromosomes of the eggs and the sperm. They are the only link between one generation and the next. We have also known for years that no blood passes from the mother to the fetus, because the blood cells of the mother are far too large to be able to pass through the placenta, and so are the blood cells of the fetus. Fetuses manufacture their own blood, and the characters of its various blood cells are demonstrably different, both structurally and functionally, from those of either of its parents.

These facts should have forever disposed of the ancient notion that the blood of the mother is continuous with that of the child. However, the myth that blood is the vehicle of heredity persists. It persists because, although today the public at large hears or reads about genes, DNA and genetic engineering, it has little understanding of genetics. Few high school graduates have been taught any meaningful concepts about heredity. Most college students never have had a course in genetics; if they have, it was probably an introduction, of which they remember only the famous Mendelian 3:1 ratio in peas and the concept of dominance, which they falsely believe applies to all characteristics.

The principles of genetics are far harder to grasp than those of the blood theory of inheritance. So one can understand why people cling to the latter. It is far easier to talk in a general way about parental characteristics being transmitted to and being mixed into a child than to refer to the segregation and recombination of genes and to their expression as characters in the child.

Thus, people continue to use such expressions as: He is of my blood; blood will tell; blood is thicker than water; half-blood, mixed-blood, and so on. We often hear that a person acted a particular way because of his or her Irish blood (or some other type of blood). Though the main culprits in keeping these myths alive are journalists, novelists and television producers,[5] we still find biologists using such expressions. For instance, Richard Leakey[6] says in one of his recent books, "Fossil hunting was in my blood." And one biology teacher interviewed in *Science*, the official publication of the American Association of Scientists, talked about "dilute Indian blood."[7]

If this usage of the word "blood" expressed mainly a lack of precision of everyday speech, little harm would result. But the situation is more serious than that. People believe and act as if blood were actually the carrier of heredity. And as long as they do, they will not only inevitably misunderstand the many problems in which heredity plays a part, but also fail to realize the economic and human consequences of such a myth. For example, an unfortunate consequence of the belief that blood is the carrier of heredity has to do with animal breeding. Even today, there are breeders who believe in telegony. Telegony is the theory that once a female has mated with a male, no matter how many other males she subsequently mates with, the heredity of her offspring by these later males will all be affected by the heredity of the first male with whom she mated. This idea sprang from the belief that the first male's hereditary qualities entered into the female's blood stream and influenced the hereditary makeup of her offspring by subsequent sires. I have known of two animal breeders in northern California who got rid of pedigreed mares because they had mated with a draft horse and had given birth to young. The breeders were absolutely convinced that these mares could never thereafter give birth to purebred offspring of their own kind.[8] Telegony has been disproved by innumerable experiments and by observing the results of matings among humans.

However, one of the most disheartening and cruel consequences of the belief that blood was the carrier of heredity was the "one-drop" rule which was used for centuries in determining people's ancestry. According to the blood theory of inheritance, as I mentioned previously, the blood of the parents was blended together to form the child; therefore, there was always a little of the blood of any ancestor flowing in one's veins. If the ancestor were considered to be inferior in any respect, it was thought that his or her blood had tainted all of his or her descendants. For example, in Medieval Europe, people considered a person to be a Jew if he or she had a Jewish ancestor as far as six generations back. In the United States, people consider a person to be black, regardless of skin color, if he or she has a single black ancestor, no matter how far back that ancestor may have been. The "one-drop" rule persisted throughout World War II. It was this rule that the Nazis used to exterminate the Jews. To them, having one Jewish grandparent was enough to classify someone as Jewish and have him or her exterminated. It made no difference what the religions of the other grandparents were or what the religion of the individual was; the Nazis believed the blood of a Jewish ancestor tainted the victim. It was this idea that also led German authorities to prevent blood transfusion from Jews to non-Jews. Jewish physicians were reported to have been sent to concentration camps for having committed such a "crime." In the mind of the Nazis, the physicians who did this had obviously tainted the blood of "Aryan" people.

The same idea was also prominent in the United States. During World War II, forty years after the birth of the science of genetics, the American Red Cross segregated the blood given by blacks from that given by the rest of the population because it was feared that through blood transfusions characteristics such as skin color from "Negroes" would be transferred to "non-negroes." If it was all right to transfuse blood between "white" GI's or between "black" GI's, it was not all right to do so between "black" and "white" GI's, and vice versa. One wonders how many GIs, "black" or "white," died in Europe and in the Pacific because they were prevented from getting the right blood type from donors whose skin color did not match theirs.

Today, although blood transfusions are given regardless of skin color, indicating that we have made some progress in understanding how traits are transmitted, we have made very little progress in understanding that "race," being an artificial classification based on specific traits, cannot be inherited. I want to emphasize again that people may inherit the genes for blue eyes, or blond hair, but they cannot inherit the genes for the "white" race.

However, the idea of blood inheritance is still with us, in particular in the domain of race. For example, for years children of a couple of which one is "black" and the other "white" or of an American Indian and "white" couple, have been called "half-breeds" or "half-bloods." A half-blood is supposed to be half of one race and half of another. But, because societies had, and still have, a strange need to classify citizens, they have generally decided that "half-bloods" belong to one race only. A child of a "black" and "white" couple is classified as "white" in Brazil and as "colored" in South Africa, but this same individual is classified as "black" in the United States. A child of a "white" and American Indian couple is often considered to be an Indian.

This way of classifying people is a consequence of the "one-drop" rule mentioned previously. It is still with us today. A fictitious story, but that one could be true, was presented a few years ago on a television miniseries, where, in a case of a divorce and child custody, a child of a Navajo father and a "white" mother was considered a Navajo. Neither the judge nor the lawyer for the mother raise the question of why the child should not be considered "white." After all, the mother was "white." Or did everyone concerned with the case believe in animalculism, an eighteenth century theory of inheritance which assumed that every characteristic of the child was attributed to the father and none to the mother?

Neither in the above case nor in the following one, a true story that appeared on the CBS television program *60 Minutes* a few years ago, did anyone raise the question of whether "race" itself was an inheritable trait. The story related to a child, born on the Navajo reservation in Arizona, who was adopted at the age of three by a "white" couple in Ogden, Utah. When he was ten years old, his biological mother wanted him back. She was supported in her quest by the Navajo tribe. The adoptive parents, the biological mother, the boy and the mother's lawyer were all interviewed by CBS. The adoptive parents argued that a woman who had given her child for adoption could not ask that it be returned to her. The boy did not want to live on the reservation. The way he talked indicated that he loved his adoptive parents and wanted to continue to share his life with them. He also wanted to continue playing baseball with his friends, a sport at which he excelled. The biological mother wanted her son back because she said she had made a mistake in having her son adopted. Her lawyer argued that no Navajo child could be adopted by a white family without the consent of the tribe. Since the tribe did not give its consent, the child had to be returned to his biological mother. The lawyer also argued that the child had to be brought up in the Navajo culture because he was an Indian, adding that this was part of his heritage. This argument should have been discussed and even rejected since social inheritance, of which culture is a part, is not the same as biological inheritance. But no one during the television interview ever raised that question. No one ever raised that question in court when, a few months later, the court ruled in favor of the adoptive parents.

All these cases, and we could find many others, are examples of how a means of looking at people which has no scientific basis affects our lives. But there is hope that things might change. For example, one finds sometimes unexpected statements by individuals that reflect a deep understanding of why "race" in humans and blood inheritance are myths. In 1951, Judge Warring, of the Fourth Circuit Court of Appeals in Charleston, South Carolina, in his dissenting opinion on the Briggs School Segregation case, wrote the following: [9]

> The whole discussion of race and ancestry has been intermingled with sophistry and prejudice. What possible definition can be found for the so-called white race, Negro race or other races? Who is to decide and what is the test? For years, there was talk of blood and taint of blood. Science tells us there are but four kinds of blood: A, B, AB and O and these are found in Europeans, Asiatics, Africans, Americans and others.[10] And so we need not further consider the irresponsible and baseless references to preservation of "Caucasian blood." So then, what test are we going to use in opening our school doors and labeling them "white" or "Negro"? The law of South Carolina considers a person of one-eighth African ancestry to be a Negro. Why this proportion? Why not one-sixteenth? I am of the opinion that all of the legal guideposts, expert testimony, common sense and reason point unerringly to the conclusion that the system of segregation in education adopted and practiced in the State of South Carolina must go now. Segregation is *per se* inequality.

Here was a judge, not a biologist, who had recognized that the blood theory was complete nonsense, and that race could not be inherited or even defined. He made the points made in this chapter and applied them to a specific problem. Unfortunately, since his opinion was a dissenting one, it had no effect on the final decision in the case in question.

Notes for Chapter Four

1   Science takes its words from ordinary language, but in doing so it often alters their meanings. As the same thing happens, too, in other branches of knowledge, it is inevitable that very different meanings come to be attached to a word, according to the way in which it is used and the person using it. The word "inheritance" is a case in point. In law we talk about the inheritance of money, land, or other property that can be taken away from us. In the same way, we can inherit customs from our parents, but we can modify them or even abandon them. In biology, however, we cannot get rid of the characteristics that we "inherited" from our parents, such as blue eyes or baldness.

2.   I have said previously that according to the blood theory of inheritance, a person is a mixture of the blood of all of his ancestors. But there is also another fallacious theory that people believe to be true. This is the jig-saw theory, according to which a person is made up of parts of such-and-such ancestral stocks in given amounts. For example: one-fourth English, one-fourth Irish, one-eighth German, one-sixteenth Italian, etc. But, in fact, ancestry cannot be broken down into identifiable ethnic fractions. One does not inherit German blood or Italian blood. One inherits genes that lead to blue-eyes, freckles, or other traits. Beyond one's parents, one can only guess at the derivation of one's ancestry in terms of genes and chromosomes. This is what the science of genetics teaches us. This important idea will be developed in a following chapter.

3.   The table is modified from one that appears in Robert Olby's, *Origins of Mendelism*, 2nd ed. (Chicago: The University of Chicago Press, 1985).

4.   See Morsink Johannes. *Aristotle on the Generation of Animals: A Philosophical Study*. (Washington DC: University Press of America, 1982).

5.   The commentator on the TV program *Nature* spoke of "Celtic and Norse Blood" as he commented on the history of the "Hebrides Islands" in the spring of 1993.

6.   Richard Leakey and Roger Lewin. *Origins Reconsidered: In Search of What Makes Us Human*. (New York: Doubleday, 1992).

7.   *Science*, November 13, 1992, 1231.

8. The idea of telegony was still prevalent in Germany. In 1919 an anti-semitic novel was published by Arthur Dinter, entitled "The sin against the Blood," in which it is argued that the blood of the German people was being "poisoned" by the aftereffects of mating with Jews. This was because the semen of a man of an alien race is absorbed immediately and completely into the blood of the female in intercourse. Therefore, a single contact between a Jew and a woman of "another race" is sufficient to corrupt her blood for ever. With his alien seed she also acquires his alien soul. She can never again, even if she marries an "Aryan" man, bear "Aryan" children. The idea of telogeny can be found also among the lower classes in Japan. It is believed that if Japanese women gave birth to a black baby whose father was a "Negro," the next baby whose father was Japanese, would show some black tinge on the body. In other words, in the minds of these people (But I am sure this is also true in other nations) impregnation of a Japanese woman by a Negro man was associated with blackening of her womb as though by ink so that the second and even the third baby conceived in it would be stained.

9. Carl T. Rowan, *Dream Makers and Breakers: The World of Justice Thurgood Marshall* (Boston: Little, Brown and Co., 1993).

10. At the time there was only one known human blood group, the ABO group. Today, we know of at least 60 human blood groups.

Chapter 5

RACIAL TRAITS: MORE FICTION THAN FACT

It was my experience in Africa among a single population located in a "typically Negroid" area of the Ivory Coast that the amount of variation in skin color, nose form and hair color was, to my originally naive eyes, rather startling. Skin color varied from very light brown to dark brown (I have yet to see a 'black' African). Nose form varied all the way from broad and flat to aquiline, and I even found occasional redheads and dusty blondes in the population. The latter trait is often missed by physical anthropologists, who spend only a short time in contact with groups they study. Blondism is considered ugly in West Africa and most people who have hair which deviates from the preferred black simply dye it. [1]

Alexander Alland

Alexander Alland,[2] like many who are not familiar with African populations, had a stereotypical view of an African as a man or a woman with a dark skin, a jaw that projects forward, curly black hair, a flat nose, slight chin, thick lips and small brows. His view was shattered when he had the opportunity to live for some time among the members of a particular African tribe. There, contrary to what he expected, he found that not all of the members of that population had similar skin color, hair color, eye color, and hair form. He became aware of the tremendous human variability that existed within this small population. The reason for his astonishment was that he, like most of us, had been misled into believing that all the individuals of a particular population have traits that are present in combinations peculiar to their group. These highly visible physical traits have been called racial traits.

At first glance, the idea that each human population is characterized by a unique combination of specific traits seems to make sense. It is not hard to distinguish a dark-skinned African from an Australian aborigine or a dark-skinned Indian because these people *tend* to have some traits in association that the others do not. Other peoples tend to have some traits in association that others do not have. For instance, Europeans *tend* to have light skin, straight or wavy hair, noses of narrow to medium width. Sub-Saharan Africans *tend* to have dark brown or black skin, wiry hair, and so on. A third grouping of traits occurs with high frequency among East Asians. Here, most people have pale-brown to slightly "yellowish" skin, straight black hair, and dark brown eyes. However, a critical analysis of human diversity around the world reveals that this generalized "racial" view is, at best, simplistic.

First, as mentioned in chapter 2, if we can perceive several main groups, there are millions of people who cannot be pigeonholed into them because they have one characteristic of one group and another characteristic of another group. Second, there is extreme diversity among individuals of the same group. For example, not all Africans have dark skin. Not all the Asians have skin flaps (scientifically called epicanthic folds) over their eyes. Third, no "racial" trait is restricted to one specific human group. For example, take skin color. There are many individuals in the world with dark skin. Some live in Africa, others in Australia, and still others in India. Take another trait, the epicanthic fold. It is very frequent among the populations of East Asia, but it is also frequent among some native dark-skinned inhabitants of southern Africa. It also occurs in some European children, but disappears once they are adults. This common trait is part of our human heritage and is expressed more strongly in some individuals than in others.

However, the most glaring problem in "race" categorization is that not all the individuals in a particular group have the combination of traits that they are supposed to have. For example, there are many dark-skinned Africans who do not have thick lips, whose hair is not thick and curly, and whose noses are not broad. There are many individuals in Asia with epicanthic folds who do not have small noses and/or medium lips.

Precise observation leads us to discover that black eyes do not always go with dark skin or even with black hair. Light-colored eyes appear not only in western Europe, where they are common, but sometimes among American Indians and Africans. Blond hair with heavily pigmented skin occurs among the Australian aborigines of central Australia, who refer to it as tawny hair and believe that a blond native Australian is a reincarnated god who desires to pass some time on the Earth.

Hence, to associate one particular trait with another, or even several, as is usually done in race classification, is not appropriate, since there are too many exceptions to this approach. The fact is that it is only in a very loose way that the traits which have been used to classify mankind into races are associated with one another.[3] This is indicated by the fact that, when anthropologists have drawn maps of the incidence of a particular trait around the world, each single trait has an independent distribution from any other. In other words, if you try to establish categories of human beings on the basis of one trait, say, skin color, certain broad divisions do seem to emerge. If, however, you use another trait, say, hair shape, you get other divisions that overlap with the skin color divisions. The picture gets worse, not better, with every additional trait that is mapped. Simply, traits such as the color of our skin, the shape of our jaw or of our hair, and the thickness of our lips are largely independent of one another. They do not cluster to form a particular "racial" type.

To believe in "racial types" is to believe that characteristics such as skin color, jaw shape, hair texture, chin dimension, lip configuration, brow shape are inherited together. But in fact they are not. Children of parents, one parent being dark skinned with thick wavy hair, the other being light skinned with straight hair, might have any combination of these characteristics.

Though "racial" traits such as skin color, shape of hair or eyes and so on, are not inherited together, they are still inherited. However, recently it has become increasingly clear that a characteristic found in one specific human population and not in another may not necessarily be due to heredity, but may have been produced by the culture or the environment in which the population lives.

For example, throughout most of the world, people have the upper incisors over the lower ones when their mouths are closed. This is called an overbite. However, it had been reported for many years that among Eskimos, the incisors met edge to edge in the closed mouth. This was long assumed to be a "racial" feature of Eskimos. It was, therefore, a shock to anthropologists to discover that young Eskimos in the twentieth century had an overbite. The overbite is a very recent development. The edge-to-edge bite was common among our remote ancestors and persisted in England until the eleventh century. It seems, therefore, that the difference between the overbite and the edge-to-edge bite is not hereditary at all but due entirely to the way our teeth are used while they are developing. Hard chewing reduces the projecting cups on baby teeth so that the mouth can be comfortably closed in many positions. If the front teeth are used to grasp food while cutting, the new molars, when they erupt from the gums, will settle into a position that allows them to meet edge to edge. On the other hand, if the food is soft, the cups of the baby teeth will not wear down and the upper incisors will grow down over the lower ones, forcing them back.

One wonders how many other "racial" traits are the result of environmental influences. For example, height, a noticeable physical feature, has been considered to be a racial trait, because some human groups are taller than others. For example, the pigmies in Central Africa are, on average, four-and-one-half feet tall, whereas Nilotic males of East Africa are six-and-one-half feet. It is highly possible that a large genetic component accounts for these differences in height; but one should be cautious about making such generalizations. Every human group seems to have its tall and short people. Height should not be considered a "racial" trait for it undoubtedly has a very large environmental component, as witnessed by the fact that in recent years there has been a rapid increase in average stature all over the world. This rapid change which occurred in two or three generations, was most likely brought on by improved nutrition and acquired immunity to diseases during childhood by vaccination or other means.

We have seen, thus far, that three assumptions upon which race thinking is based are contradicted by observations that any impartial observer could make. It is not true that human beings can be classified into groups without any ambiguity. It is equally untrue that physical characteristics by which these so-called groups are classified are transmitted "through the blood." Finally, it is untrue that these physical characteristics are transmitted as clusters from one generation to the next.

Notes to Chapter Five

1. This reminds me of an incident that occurred to me a few years ago when I was still teaching genetics. There was in the class a very dark-skinned young woman with curly hair, which reminded me of my own when I was a child. Calling attention to her hair, I explained the inheritance of hair shape. The young woman did not say a word, but two weeks later showed up in my office with straight hair—her own. She was so much changed that I did not recognize her until she told me that she often wore wigs, preferably ones with curly hair.

2. Alexander Alland, *Human Diversity* (New York: Columbia University Press, 1971).

3. However, the fact remains that many Africans have woolly hair and dark skin. Likewise, many Asians have eyefolds, dry ear wax and shoveled teeth. Why do these characteristics appear "linked?" Because in the past—and it is still true to some degree today—these populations were isolated, or rather semi-isolated, and the chances that they came in contact and interbred with populations which did not have their characteristics were small. However, as human populations disperse and intermix—a tendency in this modern world—these "racial traits" will break up. However, this does not mean that appearances of people will become uniform, but rather that "racial traits" will not be found together in the same individual as often as they are today.

Chapter 6

YOU CANNOT JUDGE A BOOK BY ITS COVER

Like most Americans who were white, I did not know what a black athlete was like. I just assumed they were not good enough for the big leagues. I had heard the talk, you know, that if you threw at them, they backed down.

Pee Wee Reese

This is what Pee Wee Reese said[1] when he learned that, on August 28, 1945, Jackie Robinson had signed a contract with Branch Rickey, which led him two years later to be the first black to play major league baseball. Reese, like most "white" people at the time, believed that "blacks" were athletically inferior to them. Sure, they could box, so the argument went. After all, there was Joe Louis, the "Brown Bomber." They *could* play baseball, there were "colored" baseball teams; but they were not as good as "whites". They could play basketball; after all, there were the Harlem Globetrotters, but, though they were excellent, they were better known as clowns. Since no black face was ever seen in professional tennis matches, in ice arenas, or in swimming competitions, it was assumed that "blacks" could not master these sports. Many white people also believed that blacks could not fly airplanes. This assumption was severely challenged when Eleanor Roosevelt took a plane ride with a "Negro" pilot during World War II. The subsequent experiences of the now-famous Tuskegee Airmen debunked this piece of mythology.

The history of sports in the United States since the 1930s clearly demonstrates that the assumption that "non-whites" were inferior in sports was unfounded. The breakthrough in track and field really began during the 1936 Olympics, when Jesse Owens broke world records in the 100-meter and 200-meter dashes and the broad jump. Owens, a triple gold medalist, did more. He destroyed the Nazi myth that the "tall, blond, blue-eyed Aryans" were supermen.[2] In the 1950s the "color barrier" also was broken in professional basketball when Walter Brown drafted Chuck Cooper to play for the Boston Celtics.[3] The color barrier in football was finally broken in the 1960s, although it is true that as early as 1899, several "black" athletes played college football at the University of Massachusetts and at the University of Michigan. These instances, however, were exceptions.

For years southern college teams refused to play their northern teams with integrated squads. Usually, black players were benched when their schools played in the South. Such policies frequently resulted in defeats. A major change came with the end of World War II, when a wave of blacks came into American colleges with benefits from the GI Bill. Many of them played sports. College football was definitely integrated. It took more time for professional football to do the same.

The fact that, today, "black" athletes dominate professional basketball, has led many critics to reverse their opinions about black athletic ability. Sports commentators, such as Jimmy (The Greek) Snyder, have declared without proof that blacks are genetically better basketball players than whites. The trouble with this type of reasoning is that, since only "white" faces were seen in ice skating events, tennis matches, or in swimming and golf (that is, until recently), many people were led to conclude that, in these sports at least, white athletes were superior. However, it is hard to believe that, even if they did exist, that *black athletic genes* would function better in some sports than in others. The truth is that blacks dominate basketball not because they are taller, have bigger hands, jump higher, or run faster than whites, but because of a whole series of complex social and economic reasons. Baseball, basketball, and football are professional sports that are financially rewarding; swimming is not. And though today professional golf and skating are rewarding, it costs a lot of money to become proficient in them. Hence for years only affluent whites could afford to pay for the years of training necessary.

Jimmy the Greek had an explanation for the fact that there were no "blacks" in swimming competitions. According to him, the bone density of blacks is greater than that of whites and because of this, blacks were not "buoyant enough." The trouble with his explanation is that the difference in bone buoyancy between blacks and whites is far too slight, if it exists at all, to make any difference.

Of course athletic ability is not dependent on the color of one's skin, the shape of one's nose or eyes, or the texture of one's hair. Nor is musical ability any more dependent upon these things. Yet, the idea that certain kinds of people are more fit than others to play or sing specific kinds of music, or to dance specific dances, has become ingrained in our ways of thinking that we are astonished to discover that non-Italian or non-Spanish tenors can sing O Sole Mio marvelously, that other than German or Austrian pianists can play with virtuosity Beethoven or Mozart, that non-South Americans can, with ease, dance rumbas and tangos.

A baritone, named Jaijung Fu, is very much at ease singing Italian arias with Luciano Pavarotti. A superb pianist, Mitsuoko Uchida, thrills concert-going audiences with his Mozart sonatas and concertos. Classical pianist André Watts is critically acclaimed and he is considered black. Trumpeter Wynton Marsalis is as much at home playing classical scores as he is jazz. There is a country music star, Shodji Tabuchi, who was trained as a classical violinist in Japan and who now plays the fiddle in Tennessee.

Today, non-whites play tennis, skate and play golf. Malivai Washington and many other fine black athletes play professional tennis in the shadow of Arthur Ashe.[5] Among the female skaters in the Olympics are Surya Bonali and Kristi Yamaguchi. Neither resembles Sonja Henie. "Black" skaters now also play hockey; for example, we can see Grant Fuhr playing goalie for the Buffalo Sabers and Dale Craigwell landing slapshots for the San Jose Sharks.

We had in this century a man who was not only a fantastic athlete but also a superb artist. His name was Paul Robeson. He was not only an All-American football player but an internationally acclaimed singer, stage actor, and film star. He was not the first black All-American football player. Fritz Pollard was. Paul Robeson was not the first black actor hailed for his talent by the white press. Ira Alridge was. Robeson was not the first black singer to fill concert halls. Roland Hayes was. But Robeson combined their gifts. His talents were exceptional. Not only did he excel in every sport he tried, not only did he have a beautiful baritone voice, but he was an outstanding student. He was valedictorian of his graduating class at Rutgers College. He had a degree in law. He had a linguistic ability that enabled him to communicate in several languages.

Today, we have become aware of the *physical ability* of non-whites in sports because of their high visibility. We have become aware also of their musical abilities because of the high visibility of the performers on the stage. However, whites tend to be less aware of the literary and scientific abilities of non-whites.

For instance, the beginnings of American black literature are found in the slave narratives of Gustavus Vassa[5] and Frederick Douglass. [6] Written by slaves, or former slaves, these eloquent documents belied claims of black inferiority by slavery advocates. The slave narratives were also produced after slavery was officially abolished. They were autobiographies of individuals who refused to accept the denial of their individuality and of the realization of their own aspirations, which the white world demanded of them.

Richard Wright's autobiography,[7] for example, is the story of a man who refused to be subservient to American racial norms and was willing to rebel to the point of self-annihilation in order to remain inviolate. Since, for him as a black man, writing lay outside the structure of socially acceptable possibilities, it was a means of expressing his individuality, of self-assertion and rebellion.

The first published black American poet seems to have been Jupiter Hammon. Born as a slave in 1711, Hammon was concerned more with things in heaven than on earth. However, the best known early black poet was Phillis Wheatley. Born in Africa and transported as a child-slave to America in 1761, she was employed by a Boston family who gave her some education. The oldest daughter of the family, Mary, was designated Phillis's tutor, and within sixteen months Wheatley had mastered English to the point that she could read the most difficult parts of the Bible. Within a brief time she began writing poetry. [8]

Frederick Douglass also learned to read and write at an early age. Few slaves had the opportunity. However, to become writers and poets, as Phillis Wheatley and Frederick Douglass did, blacks needed to know more than how to read and write. They had among other things to be gifted individuals. Both Wheatley and Douglass were.

Unlike Phillis Wheatley, who was fortunate to grow up among compassionate individuals, many of the black writers, including Frederick Douglass, had to overcome a childhood lived in poverty, abuse, racism, loneliness, and often homelessness. This was the case of Richard Wright, "who developed from an uneducated lonely Southern boy to become one of the most cosmopolitan, well read and politically active writers in the history of American literature."[9] As time passed, the number of black writers increased. One significant development was the degree to which young black fiction writers were able to obtain relatively high levels of formal education and therefore establish a firmer grounding in contemporary literary models. Black writers have received the highest accolades. The first black Nobel laureate was a man born in Nigeria, Wole Soyinka. The second one was an American woman, Toni Morrison.

If today most whites are aware that blacks can be excellent athletes, musicians and artists, and that blacks write, they are not generally as conscious of black inventors and scientists. How many know that William H. Richardson improved the baby carriage by making it possible to change the direction of the carriage without actually turning the carriage around and disturbing the baby, or that Jerry Certain improved the parcel-carrier for bicycles to help make it safer to carry packages while riding. How many of us know that David A. Fisher improved the furniture caster by inventing a new type of spindle so that the casters were firmly held in place and would not easily come off when the furniture was moved? How many of us know that, in order to increase the safety of the trolley car, Elbert R. Robinson improved the construction of the contact arm on top of the car to help prevent it from jumping from the wire when the car rounded a curve or going down an incline in the road. How many of us know that the first pocket fountain pen was invented by William B. Purvis and that Joseph H. Dickinson invented the roll of perforated sheet music which automatically enabled the mechanisms in the piano to play a melody. Dickinson also improved the sound quality of the record player by inventing the tone arm, which gave a richer tone to the record, and the horn that increased the volume so that the music could be heard from a greater distance.[10].

Perhaps the most famous black scientist was George Washington Carver, the first "colored" graduate of Iowa State College. For a long time he did not know whether he wanted to be a painter or a scientist. Fortunately for agriculture he became an excellent scientist. Like so many of the "black" writers, Carver had to overcome a childhood of poverty, racism, loneliness, and often homelessness. He was born a slave in 1860 in the State of Missouri, three years before emancipation. He lost both parents during the Civil War when he was still a baby and was raised by Moses and Susan Carver, an elderly couple who had owned little George's mother. The first few years of his life were somewhat secure and safe, living and working on the Carver's farm. However, George had a yearning within him to learn about nature, especially about plants. He wanted to know how roses became double, why not all the leaves on a tree were the same, why clover and oxalis folded their leaves at night and on dark days, what insects were doing in the flowers. This consuming desire to know pushed Carver to learn how to read. As have many outstanding children who have had access to a few books, he managed to do this by himself.

At the age of ten, Carver abandoned the security of home and the safety that lies with familiar things and set out to find a school that would open its door to him. He not only found one but others that led him onward to higher education. After graduating from Iowa State college, his love and knowledge of plants led to his being offered a position as an assistant botanist in the experiment station attached to the College. In a few years he became one of the best-trained agriculturists in the country. At the age of twenty-three, he joined the Agricultural Department at the Tuskegee Normal and Industrial Institute.

Cotton was the main crop then in Alabama, and Professor Carver was able to help increase its yield, not only by better management practices, but also by using better varieties that he himself had developed by hybridization. These varieties were in general more resistant to diseases and better adapted to the light, sandy soil conditions of the Southern States. However, Carver became well aware of the harm done to Southern agriculture by having only one crop, cotton. The cotton plant is a heavy feeder which, grown year after year, depletes the soil of its nutrients. The dangers of a one-crop system from an economic as well as an agricultural point of view had been denounced for years. One way to solve this problem was to rotate crops: one year, cotton; the next year another crop; the third year another crop. Then, the three-year cycle is started again. Carver persuaded the farmers to plant peas, soybeans and peanuts, plants that do not require as much fertilizer as cotton because they can extract nitrogen from the air and impart it to the soil, instead of using up the nitrogen in the soil.

George Carver was the man responsible for making the soybean a field-crop in the United States. This plant, a very important source of food for the Chinese, had been brought back by Commodore Matthew C. Perry in 1854, but nothing had been done with it. Today, soybeans are one of the most important sources of vegetable oil and protein. When George Washington Carver died at the age of eighty-three, he had been recognized not only as a great scientist, but also as a benefactor of mankind, as a man who demonstrated that in human ability there is no color line. [11]

Another famous "black" scientist who, like Carver, beat the odds overcoming poverty and racism was Dr. Ernest E. Just, who as a child moved from Charleston, South Carolina, to New York in search of better schools. He found them and graduated magna cum laude from Dartmouth College and obtained his Ph.D. from the University of

Chicago. If Carver was a scientist whose discoveries were practical and understood by many, Just was a theoretical scientist whose discoveries pertaining to the biology of cells and fertilization of sex cells were understood by only a few.

Carver and Just were exceptional people, not just because they were outstanding scientists but because they were academics. Most blacks who had any scientific ability pursued medicine as a career. None of them could be blamed, for they would have been slightly out of their minds to battle for position and eke out a meager existence in the academic world when jobs and money were to be had so readily in medicine. A few of these physicians became famous surgeons and scientists. Among them was Daniel Hale Williams, who, in 1893, under the most difficult conditions, performed the first open heart surgery, and Charles R. Drew, who is remembered today for pioneering the development of blood preservation.

It is clear from the above examples that skin color, shape of nose or hair, have nothing to do with physical or intellectual abilities. Today we have athletes from every part of the world, of very diverse physical appearance, competing in every Olympic event. This human diversity is also reflected in the scientific and artistic worlds. Yet, we still have a tendency to associate physical appearance with modes of behavior, such as manner of speech. For example, we are astonished to find someone who does not look at all like Jack Kennedy, but more like Yasir Arafat, speaking English with a New England accent. An old friend of mine told me recently that he was taking a nap in his favorite chair in front of the television which was on. As he woke up, he heard, but did not see, who was talking. My friend was impressed by what the man was saying and how he said it. I guess he was expecting Walter Cronkite, but as he opened his eyes he was astonished to learn that the man on the screen was Ed Bradley, who is considered a black man.

Another example of similar misguided association involves Asian Americans, who tell the following story in varying detail but is nonetheless is the same story. In a speech given at Michigan State University in the Spring of 1994, Ronald Takaki, author and professor at the University of California at Berkeley, told the audience that a few months before, he was in a taxi going to give a lecture at a South Carolina university, and the taxi driver told him: "You speak very well English. How long you have been in this country?" The taxi driver did not realize that many Asian Americans, like Professor Takaki, were born in the United States from parents who were themselves born in

America. That they speak English, and in some cases excellent English, derives from the fact that they went to grammar school in the United States. This appears also to be the case of Judge Ito, who presided in the O. J. Simpson case. If we look at the way a child acquires a language, we can see why this is so.

Children begin to speak single words at about one year of age; by age two they start joining words into two-element phrases. At age four they have mastered the fundamentals of grammar and speak in sentences. This ability is that part of language skills that pertains to heredity. Which language a child will speak depends upon the environment in which he or she is raised. A child born to English-speaking parents and raised in their household develops an ability to speak English. But the same child, placed at the age of six months into a French household in Paris and raised there, would never utter an English word and at four would be speaking as a four-year-old French child with no detectable English accent. This is because, by the age of four and five, a child has usually mastered the fundamentals of the language to which he or she has been exposed.

Children hearing two languages usually learn both without confusing them. Up to the age of ten or eleven a child can assimilate a second or third language with much the same ease that he or she learned the first, speaking without an foreign accent and without grammatical clumsiness. From the age of twelve on, learning another language not only becomes progressively harder but the very process of learning changes. The unconscious assimilation characteristic of the young child is replaced by a conscious effort. Pronunciation becomes a task, vocabulary rote learning, and grammar the mastering of a new set of rules. The immigrant who comes to a new country before the age of eleven learns to speak the new language like a native. As a young adult, he or she will not be recognizable as a foreigner, either by accent or by speech pattern. But the immigrant of eighteen or nineteen will probably never speak without a detectable accent. One who arrives in his mid-twenties, as I did, is likely to retain a strong accent and never be completely at home in his new language, especially with regard to the use of prepositions, the subtleties of word order, and the use or omission of articles. It is believed that the reason for this is that at the time of puberty, something happens to the hearing process. An adult does not hear as distinctly as a young child and therefore often cannot say a sentence the same way a native speaker says it. [12]

Mannerisms, such as movements of the mouth, eyes, hands, and head, or ways of walking, observed in some groups of people are no more inborn than their languages or accents. For example, people living around the Mediterranean Sea tend to gesticulate far more than northern Europeans. But this mannerism is a form of habit that one can easily discard. Children who have been removed from their homes early in their lives and brought up elsewhere are found to be able to lose those gestures. Adults like myself, who moved from one country to another have been known to decrease these mannerisms but not lose them entirely.

The best example I know of to show that there is no link between physical appearance and language and behavior in general is the following. [13] There were two friends who lived in Manhattan, New York City. One, named Paul, was "Chinese" looking; the other, named Joseph, was "European" looking. The strange thing about these men was that Paul, who was born in New York, did not know Chinese and never used chopsticks, but his friend Joe, who was adopted by a Chinese couple who moved from New York to Canton, was raised in China, spoke Chinese and wore traditional Chinese dress. When World War II broke out, Paul found himself in a Chinese American unit in which he was a perfect outcast, since he knew nothing of the Chinese customs. In a similar fashion, Joe found himself in a "white" regular army unit in which he was a perfect stranger. The Chinese was Joe, not Paul, who was American in all but appearance. This example, in which two men's natural environments were reversed, shows that individuals are not born with a specific culture. On the contrary, they acquired their culture during their childhood, just as they learned specific languages. [14] This obvious truism is not understood among many social workers, who believe that "black" or "American Indian" children should not be adopted by "white" couples because such an adoption would work against the children's "racial-cultural" inheritance. These workers do not realize that they are perpetuating racism. Their actions help to keep alive the myth that the physical appearance of an individual reflects a special mindset. Such a myth is unfortunately well distributed among people. It has led to cruelty, even to killing as happened a few years ago in Detroit when an "Oriental looking" man was clubbed to death because he symbolized the American-Japanese rivalry in the automobile industry. The man who got killed was in fact not a Japanese, nor a Japanese American, but a Chinese American who worked for the Ford Motor Company.

Notes:

1. Harvey Frommer. *Rickey and Robinson: The Men Who Broke Baseball's Color Barrier* (New York: Macmillan Publishing Co., 1982).

2. Jesse Owens (with Paul Neimark), *Jessie: A Spiritual Autobiography* (Plainfield, N.J.: Logos International, 1978).

3. Neil D. Isaacs. *All the Moves. A History of College Basketball.* (Philadelphia: J.B. Lippincot Co., 1975).

4. When Arthur Ashe died on February 6, 1993, he was hailed as a hero for having been not only a world famous tennis player, but also the first black man to win Wimbledon and the U.S. Open. The last statement as it stands is open to the following query, why is it astonishing that a black man can be an international tennis champion? Is it because he is physically different from a white man? Or is it, because, being "black," he did not have the same opportunities to play tennis that a white man and therefore it was harder for him to become a champion? In the case of Ashe the last reason is correct, since he had to leave his hometown, Richmond, Virginia., because he was unable to improve his tennis game in a segregated city where he could not practice his favorite sport in public parks.

5. Gustavus Vassa. *The life of Oluaundah Equiano or Gustavus Vassa, the African,* 1789.

6. Frederick Douglass, *Narrative of the Life of Frederick Douglass: An American Slave* (1849; Cambridge, Mass.: Belknap Press, 1960).

7. Richard Wright, *Black Boy* (Cutchoge, N. Y.: Buccaneer Books, 1991).

8. Phillis Wheatley, *Poems on Various Subjects, Religious and Moral* (London, 1773).

9. Joyce Ann Joyce, "Richard Wright, 1908-1960," In *African American Writers.* Valerie Smith, Lea Bacheler and A. Walton Litz, eds. (New York: Collier Books, Mcmillan Publishing Co., 1991).

10. A list of these inventors with their respective inventions can be found in black science activity books addressed to grammar school children. They are published by Chandler/White Co. Inc., Chicago, Illinois.

11. Four years before his death, George Washington Carver received from the University of Rochester a doctorate of science degree, *honoris causa*, in recognition of his scientific work and for being a role model for thousands of individuals.

12. It is believed that this also occurs in the case of a musician. If the musician has heard music as a young child, it is easier for him or her to have a perfect pitch than

someone who familiarizes himself or herself with music past the age of twelve or thirteen.

13. Amram Scheinfeld, *Your Heredity and Environment* (Philadelphia: J.B. Lippincot. Co., 1975).

14. That a specific language is not inherited can be deduced by the fact that whole populations sometimes change language because of political or other cultural pressures. For example, the population of ancient Gaul gave up its Celtic languages and adopted Latin. A whole nation, such as Switzerland, can be bilingual or even trilingual.

INTRODUCTION TO PART TWO

In Part One, I have demonstrated the false assumptions on which race thinking is based. Yet race thinking has inhibited and undermined social cooperation and harmony by assuming that some human groups, as defined by physical appearance, are intellectually inferior to others. Hence, so their reasoning goes, people in these groups have to be treated differently, to be segregated, to receive less education, and to be economically deprived. That race thinking has played a major and unfortunate role in our dealing with social issues is well known today. But what is not so well known is that race thinking also has delayed for years the development of scientific thought. In Part Two I will demonstrate this.

Race thinking has been a hindrance to our understanding of two critical biological theories: evolution and natural selection. In spite of the fact that a massive amount of research and data existed to contradict notions that human races exist, biologists and anthropologists clung to the notion for many years. Some unconsciously, and others willfully misinterpreted the theories to fit their preconceived racist ideas.

The theory of human races, in fact, led scientists to ask inappropriate questions, which led to intellectual and scientific dead ends that proved to be costly in both time and money. For example, those who supported the biological concept of race assumed that, within particular species, distinct populations existed that could be distinguished from one another by their possession of certain distinctive hereditary traits. Based on this assumption, scientists assumed they could classify people based on combinations of these traits. But as we have seen, they were unsuccessful. The absence of success should have led them to suspect the theory of race and assumption that racial traits exist. However, they persisted in their belief, all the time asking questions about racial traits as if they were real, and as if they existed naturally in groups. For instance, does a so-called racial trait give an advantage to those who possess it in the environment in which they live? In biological terms, does a racial trait have an adaptive value? Such a question was logical, since these scientists assumed that each race was adapted to the climatic conditions under which it lived: blacks in the tropics, whites in western Europe, Asians in cold parts of the world. However, the adaptiveness of skin color, the trait upon which most racial classification has been based, has been overemphasized. Other racial traits, such as eye color, hair color, size of skull, form of hair, or shape of eyes, have never been shown to be advantageous for the individuals who possess them. In fact, most suggestions of this nature have been purely speculative. When presented systematically, these suggestions often contradict one another.

The assumption that human races exist led scientists to ask yet another inappropriate question, namely, when were human races formed? For many years the question has been disputed by anthropologists. Some believe races existed in the past before the emergence of modern human beings. Early in the evolution of that argument, some of its proponents even suggested that each of the "human races" descended from a different, distinct species of ape. Recently, others have suggested that each "race" evolved independently of the others, each from a separate strain of early man, *Homo erectus*. According to this hypothesis, the "yellow race" evolved from the *Homo erectus* of Asia, the "black" race from the *Homo erectus* of Africa, and the "white" race evolved from the *Homo erectus* of Europe.

An opposing school of thought accepts the idea that *Homo erectus* is our common ancestor, but believes that only one population of *Homo erectus* evolved into *Homo sapiens*. It was later that the branching out and forming of distinct races occurred, according to this theory.

The assumption that human races exist has played a trick on geneticists. Whereas they developed a specialized branch of scientific inquiry, population genetics, which has been a powerful force in explaining the process of race formation in plants and animals, they were not able to see that their own science led to the denial of the existence of human races. Simply stated, the conditions necessary for races to develop were not present and did not apply to human beings.

The assumption that human races exist remains a hindrance to the ways in which scientists design experiments and interpret data. Having made the assumption that there are human races—more for political, economic, and social reasons rather than for reasons based on experimentation and observation—scientists divided mankind into groups. Having established these groups, biologists and anthropologists have ignored human similarities and looked only for differences that will justify their preconceived notions of classification.

An example, which will be discussed in chapter 14, involves those who still cling to the idea that some human groups are superior in intelligence to others. They have used IQ tests to support their beliefs that these differences are real. They turned to this form of evaluation because they believed the method was more sophisticated than craniology and its crude forms of quantification. However, the results of IQ testing, seeking to link race and intelligence, have proved to be as negative as the ones used by nineteenth-century craniologists and for the same reasons. Not least among them is the faulty design of their experiments. Here we will cite just one example.

Suppose we believe that women are, in general, smaller in stature than men. Let us suppose also that we wish to confirm this notion scientifically. We take a sample of men and women from one population and simply compare their heights statistically. We are likely to find that, on average, men are indeed taller than women. If we take another sampling in the same population, *or in another,* we also will find that men are, on average, taller than women. We are able to reach this scientific conclusion because we are able to clearly distinguish men from women; there are things that are fundamentally and biologically different between men and women, namely their gender. In other words, we can make comparisons between the heights of men and women because they are two distinct groups of individuals.

However, this is not the case with the comparisons made in the experiments done with IQ testing among human groups. Consider, for example, the IQ testing done to compare the intelligence of black and white American students. To begin, how do we classify students into categories? Mostly by skin color. In doing so we will immediately run into trouble because skin color is not a characteristic that can be sharply defined. Skin pigmentation runs from very light to very dark. As a result, none of us are completely black or completely white. Thus, if you assume that there is a link between IQ and skin color, you would be far better served if you measured not only an individual's IQ, but also the amount of his or her skin color. I am not aware of anyone who has correlated skin color and IQ in this manner.

This pigeonholing approach to the classification of people has characterized many experiments that deal with human diversity, not only in intelligence testing but also in medical research, where, once again, patients are urged to self-classify or are classified into races. These assumptions affect the design of medical experiments and the interpretation of data.

Finally, the pigeonhole method is still used by the U.S. government as a mechanism for classifying its citizens. The approach has led only to confusion, contradiction, anxiety, and pain for many individuals who do not, or cannot, be forced into any of the government's dreamed up categories.

Chapter 7

DID WE EVOLVE FROM APES, AND IF SO, FROM HOW MANY?

Descended from the apes my dear, let us hope that it is not true, but if it is, let us pray that it will not become generally known.

That is what the wife of the Bishop of Worcester exclaimed when her husband told her what Professor Huxley had said at the annual meeting of the British Association for the Advancement of Science, on that fateful Saturday, June 30, 1860. The exchange occurred just seven months after the publication of Darwin's controversial but scholarly book, *The Origin of Species*. In this work, Darwin successfully organized evidence to demonstrate that living organisms had evolved and forcefully explained the process by which it took place, which he called natural selection.

If the average person in the street were asked the meaning of the word evolution,[1] he or she would likely reply (as did the wife of the Bishop of Worcester), "We came from apes." That is not, however, what evolution means. Human beings and apes are all part of the world today. We are not descended from each other. The relationship between apes and human beings is more analogous to a cousin-cousin relationship rather than that of a parent to a child. Cousins have a pair of grandparents in common. Human beings and apes are thought to have had a common ancestor in the distant past. From this common ancestor, both inherited some characteristics; thus, they resemble each other. Was this common ancestor a man or an ape? It was neither. It was a creature with such a form that it had the potential to give rise to apes on the one hand and to human beings on the other.

Biological evolution is defined as the theory that plants and animals now living are the modified descendants of somewhat different plants and animals that lived in the past. These ancestors to present-day life forms, in turn, are thought to have descended from predecessors that differed from them, and so on, step by step, back to a beginning shrouded in mystery. We can forgive the person on the street for believing that we are descended from modern apes. However, it is more difficult to forgive earlier-generation anthropologists for holding to this idea. This misconception of the process of evolution has led them to suggest that "races" of humanity have arisen from different apes, some from the chimpanzee and some from gibbons or orangutans. For example, in his book *Up from the Ape* (1931, p. 572), Earnest Albert Hooton says:

> What I do maintain is that during the Miocene period a family of giants, general anthropoid apes, the *Dryopithecus* family, which was spread over a wide zone of the Old World, evolved into the ancestors of existing and extinct forms of anthropoid apes *and those of several varieties of man* [italics added].

Under this scientific jargon Hooton has buried the implication that human races are descended from distinct ape species. Like other anthropologists of his time,[2] he accepted the notion that a link existed between the black races and the African apes, and between the Asians and orangutans, whereas European ancestry could be traced to an advanced, *more evolved* ape, which became extinct. If we were to apply Hooton's idea to the canine world, we could propose that the rough-haired Newfoundland is a codescendant of a rough-haired bear, while the smooth-haired mastiff had evolved from the sleek leopard. We would support our view by stressing the points which the Newfoundland and the bear, the mastiff and the leopard have in common, ignoring the many hundreds of characters which the two types of dogs, the Newfoundland and the mastiff, possess in common and which separate them from the bear and leopard.

Once Hooton and his associates assumed that human races descended from ape species, they asked a strange question: "Did these races evolve at different rates?" The question is strange because it seems to have been asked only about human races. No biologist has ever asked if a race of fruit flies, cows, or radishes, were "more evolved" than another. For some reason, likely to be more political than biological, it seemed important to know that Europeans had evolved more rapidly than other "races."

Whatever their reason for doing so, these anthropologists compared what they considered to be the three main races—Europeans, Africans, and Asians—anatomically. Then they described them in a stereotyped way comparing them to modern apes. If any of the so-called human races appeared to be closer in their development to apes, there would be reason to believe that they had "lagged behind" in their evolutionary development.

However, a critical analysis of this kind of work indicates that the evidence presented to support the theory was inconsistent. Ape jaws project forward to a considerable degree; their foreheads recede; their noses are broad and low. Anthropologists associated these same characteristics with Africans, less with Asians, and not at all with Europeans. There are, however, other traits that conflict with this sequence. For instance, one of the most conspicuous differences between apes and humans is the amount of body hair present. In this case, it is clearly Europeans who are the most hairy and Africans the least. Hence, in the hairiness sequence it is Ape, European, Asian, African. Apes have thin lips; Asians have thin lips, while Africans have full lips. For this characteristic the sequence is Apes, Asians, Europeans, Africans. Were we to multiply the examples, the process quickly becomes absurd. If one human group is similar to apes in certain ways, it is less similar in others, giving no evidence of evolutionary superiority for any of the groups.

The idea that the assumed human races have separate ancestries has not been abandoned.[3] There is a modern, more subtle view of it. Though anthropologists are more or less in accord that *Homo erectus*, a more primitive form of human being, was our ancestor, some believe that *Homo erectus* evolved into *Homo sapiens*, not once but at least three times. According to their theory, *Homo erectus* in China gave birth to the modern Chinese; *Homo erectus* in Africa gave birth to the modern Africans; and *Homo erectus* in Europe gave birth to the modern Europeans.

This multiregional-racial view of our origin is not supported by our knowledge of evolution. There is no example in the animal or plant world where descendants of two different populations of the same species have been isolated for a long time and subsequently evolved into the same species. If the Asian *erectus*, the African *erectus* and the European *erectus* had been isolated from one another for at least a million years, as the proponents of the multiregional racial view believe, they would not have evolved into a single species but into three different species, with the result that no member of one species could breed with a member of either of the others. The fact that any two human beings of opposite sex, however diverse, can produce fertile offspring leads us to conclude that we *all* belong to the same species and have the *same* prehuman ancestry.

To argue against this view, proponents of the multiregional hypothesis of our origin declare that the Asian *erectus* already had characteristics possessed by modern Asians, that the African *erectus* had the characteristics we see in modern Africans, and that the European *erectus* had the characteristics of modern Europeans.[4] But this "racial" interpretation of the remains of *Homo erectus* is subject to skepticism. The fossil record is inevitably limited to the observation of bones, which reveal little about how an individual looks. As W. E. Clark has commented in his book The *Fossil Evidence for Human Evolution*:

> The determination of the evolutionary differentiation of the major races of *Homo sapiens* present exceptionally difficult problems to the paleontologist. These races may be distinctive enough in the flesh, but the anatomical distinction between one and another is not reflected to anything like the same degree in the bones. While they are recognizable by a number of external (and seemingly rather superficial) characters such as skin color, hair color and nose form, they are by no means so easily distinguishable by reference to skeletal characters alone, at least not in the case of individual and isolated specimens (which are usually all that paleontology has to offer). It may indeed be possible to identify a skull from a modern Negro, an Australian aborigine or a European, in individual cases where the racial characters are exceptionally well marked; but the variation within each group is so great that skulls of each type may be found which are impossible of racial diagnosis.[5]

Since it is impossible to determine with any certainty the "racial" characteristics of our ancestors, it has been suggested that modern-day images of what the so-called "races" look like ought not to be projected back upon our remote ancestors[5]. This is a wise choice. But let us go one more step. Since it is becoming clear that human races do not exist today, let us abandon the idea that they existed in the past. After all, if it were possible to have thousands of *Homo erectus* skeletons instead of a few, we might conclude that the wide range of human diversity was already present among *erectus* individuals.

We have seen in this chapter that the belief in the existence of human races led people to misunderstand the nature of evolution. In the next chapter we will see how it led people to misunderstand and make a travesty of the theory of natural selection, the mechanism by which evolution occurs, ultimately leading to pseudoscience and to a sociopolitical theory known as Social Darwinism.

Notes:

1. Biological evolution is a theory, an attempt to explain the diversity of animals and plants. Like all scientific theories, it has been modified and undoubtedly will be revised in the light of knowledge. In its broadest sense, today's concept of biological evolution includes the following ideas:

   A. All life evolved from one or a few simple kinds of organisms.
   B. Each species, fossil or living, evolved in a single geographic location from another species that preceded it in time.
   C. Evolutionary changes were gradual and of long duration. Evolution occurs more rapidly at some times than at others. It does not proceed at the same rate among different types of organisms.
   D. Evolution can be either progressive, that is, simple forms giving rise to complex forms, or retrogressive, with the descending forms becoming less complex than their ancestors.
   E. The greater the similarity between two groups of organisms, the closer their relationship and the closer their common ancestral group in time.
   F. Over long periods of time (millions and millions of years) new genera, families, orders, classes and phylla developed through a continuation of the kind of evolution that produced new species.
   G. Evolution continues today in generally the same manner as it did in the past.

2. See the writings of Louis Bolk, Wood Jones, Herman Klaatsch and Karl Vogt.

3. The idea that different races have separate ancestries was more an attempt to demonstrate the inferiority or superiority of some forms of humanity than it was an attempt to provide an unbiased scientific explanation of human diversity.

4. It is likely that the reason why some anthropologists believe in the multi-regional origin of *Homo sapiens* is that they firmly believe in the existence of human races. If a *Homo erectus* skeleton found in China were interchanged with a skeleton found in Africa, one wonders if some anthropologists would detect the transposition.

5. W. E. Clark, *The Fossil Evidence for Human Evolution*, 2nd ed., (Chicago: University of Chicago Press, 1964), 262.

6. Marvin Harris. *Culture, People and Nature. An Introduction to General Anthropology*. 3rd ed., (New York: Harper and Row. Publishers. 1980), 98.

Chapter 8

MIRROR, MIRROR ON THE WALL, WHO'S THE FITTEST OF US ALL?

Then will the world enter upon a new stage of its history—the final competition of races for which the Anglo Saxon is being schooled . . . This powerful race will move down upon Central and South America, out upon the islands of the sea, over upon Africa and beyond. And can any doubt that the result of this competition will be the "survival of the fittest?"

Josiah Strong 1885

What Josiah Strong alluded to in the last sentence above was natural selection, the theory that Charles Darwin proposed in 1859. Notions like "survival of the fittest" and "struggle for existence" often come to mind when one is asked about the meaning of natural selection. However, neither phrase was original with Darwin. The first was used by Thomas Malthus, who, in *An Essay on the Principle of Population*, expressed the view that human reproductive capacity far exceeds the available food supply. Human beings compete among themselves for the necessities. The unrelenting competition engenders vice, misery, war and famine. The second phrase was coined by the social philosopher Herbert Spencer in 1862.[1]

If, indeed, Darwin adopted the notion of "struggle for existence," he had mixed feelings about the idea. The "struggle for existence" was a concept he used only cautiously. He must have perceived that such slogans conceal two important questions, namely, who are the fittest and what are they fittest for? We shall see that several different answers have been given to these questions. Some have disastrous political and social repercussions.

With the publication of his research, Darwin did two things: First, he convinced the scientific world that evolution had occurred, and second, he proposed the theory of natural selection[2] as its mechanism.

For this he became famous. However, he should have shared his fame with another man, Alfred Russell Wallace. Prior to Darwin's publishing his famous research, Wallace had written a short essay that he sent to Darwin in 1858. In this essay, Darwin was astonished to find that Wallace had also developed the idea of natural selection. In order to share the honor of having established the mechanism by which natural selection is brought about, Wallace and Darwin agreed to read their papers before the Linnean Society in London the same day, on July 1, 1858. However, it has been Darwin's book, *The Origin of Species*, with its impressive weight of evidence and argument that left its mark on the thinking of humanity.

Strangely, neither Darwin nor Wallace were the first to think in terms of natural selection. Darwin's own grandfather, Erasmus Darwin, expressed such an idea at the end of the eighteenth century in his book *Zoonomia* (or, *The Laws of Organic Life*). And the same idea can also be found in the writing of William Wells. In a small essay, "Account of a Female of the White Race of Humanity, Part of Whose Skin Resembles that of a Negro," Wells described certain patches of black skin on the otherwise white body of a woman named Hannah West. From his observations, he concluded first "that great heat is not indispensably necessary to render the human color black." This led him to seek his answer to the cause of dark skin elsewhere. Since Wells was a physician, it is not surprising that he thought of diseases. Suppose, he thought, that resistance to a disease prevalent in Africa was correlated in some way to darkness of skin. This would account for the spreading of black people over the continent, since those with white skin would be eliminated by the disease. Wells wrote:

> Of the accidental varieties which would occur among the first few scattered inhabitants of the middle region of Africa, some would be better fit than others to bear the diseases of the country. This race would consequently multiply, while the others would decrease, not only from their inability to sustain the attacks of disease, but from their incapacity of contending with their more vigorous neighbors. The color of this race, taken for granted from what has already been said, would be dark. But the disposition to form varieties still existing, a darker and a darker race would in the course of time occur, and as the darkest would be the best fitted for the climate, this would at length become the most prevalent, if not the only race, in the particular country in which it originated.

It is too bad that both Erasmus Darwin and William Wells wrote their ideas in a rather obscure way. They missed an opportunity to present to the world one of the most important biological concepts, natural selection, which now is attributed to Charles Darwin.

Most of us are familiar with the way new varieties of dogs or other domesticated animals are created. We select for breeding purpose only those dogs which come the closest to the type we want, removing the others from the breeding population. This type of selection is artificial, but, as Darwin pointed out years ago, nature is carrying on a similar type of selection, though not as strictly as breeders do. Most people have the following concept of natural selection. They are aware that organisms are able to produce tremendous numbers of offspring, and if all of these were to survive and reproduce, every species, even the least prolific ones, eventually would increase until they covered the earth. Obviously plant and animal populations do not actually increase in this manner, for their number is limited by the food supply and other environmental conditions that limit their expansion. In every generation only a fraction of the offspring, the best fit, survive and give rise to the generation succeeding them. On the other hand, those less well suited or damaged in some fatal way are eliminated. The struggle between those who survive and those who do not is described as direct and conscious combat, leading to a picture of nature "red in fang and claw." Such combat occasionally does occur in nature, to be sure. The picture of two stags struggling in rivalry comes to mind, and we are led to believe that the winner of the fight reproduces while the loser does not.

However, this view of natural selection as an all-or-nothing proposition, in which only the strongest survive, all others perishing, leads to a serious misunderstanding of what really happens. For example, in the case of the previously mentioned stags, the loser may well find other females, or a third stag may impregnate the doe while the combat rages between the combatants. Generally, natural selection is a peaceful process in which the concept of true combat is really irrelevant. It involves such things as better integration into the particular environment, more efficient utilization of available food, and better care of the young.

To understand what is really meant by natural selection, let us observe what happens under trees every spring. The year before, these trees produced many seeds. A large number fall on the ground. From them, numerous seedlings sprout, but, as the season progresses, the number of survivors diminishes, while the size of the ones that survive increases. Finally, there are but a few individuals left from the hundreds of seeds germinated. Among the densely populated seedlings, there has been a struggle for existence. The struggle was keen because light was absolutely necessary to every seedling. The ones that spread their leaves more rapidly and over a larger area assure their salvation, cutting off light to the others, and causing their starvation. Or if dry conditions prevail, the seedling whose roots penetrate the soil more rapidly or deeply has an advantage in survival. In both cases, the survivors go on to produce next year's crop of seeds, and the process is repeated. In this example, some individual plants are more fit to deal with their environment and hence are selected by nature. But, as with any natural phenomena, natural selection is not perfect. Some seedlings might have survived, not because they were better fit, but because the seeds from which they originated happened to fall on a piece of ground rich in nutrients and far away from the parent tree. Other seedlings that were better fit never had a chance, because they may have fallen on rocky ground or in a lake.

Overproduction of offspring and natural selection are also well known in the animal world. It is common for some types of fish to produce more than a million fertilized eggs annually. A female frog may lay as many as 12,000 eggs at a time. Most of these eggs will never become tadpoles, and natural selection may play no role whatsoever in choosing those who will. It is very possible that a bass feeding on them might consume all of them except for one or two that were overlooked.

These remnants could be individuals of any hereditary constitution, favorable or otherwise, and in this fashion, less fit individuals might attain adulthood while better adapted ones may have been killed before hatching. In the same manner, tiny newly hatched fish are indiscriminately gobbled by bigger fish, not according to special adaptive features they might develop as adults. Similarly, grazing deer and seed-eating mice do not concern themselves with the hereditary makeup of the food they consume. But in the overall picture, it is not unreasonable to assume a greater percentage of survival for better-equipped specimens and that selection is an active force in nature. And so, if there were a mirror on the wall, how would it answer the question, "who are the fittest of them all?" The ones that survive and leave the most offspring, sometimes regardless of their hereditary makeup? This answer is far from the one that is often implied in the world of business. One often hears, "Sure, I'll admit that Harry got to be a millionaire by being ruthless in business, but what's wrong with that? Our world is governed by the law of survival of the fittest." The view that natural selection exists in society as it exists in nature has been called

Social Darwinism. It equates success in business and social power with biological fitness, and equates economic laissez-faire, cut-throat competition, and rivalry with natural selection. But the accumulation of material riches or even individual survival has nothing to do with biological fitness. Suppose Harry the millionaire above, has no children, while a competitor, Joe, who he has driven into bankruptcy, has ten. Who is the fittest? It is Joe, the unsuccessful businessman, because, in true Darwinian terms, wealth and prestige are significant only so far as they affect success in leaving progeny.

A key point made by Darwin that Social Darwinists have missed, or preferred to forget, is the following. According to Darwin, the fittest organisms in an artificial selection process are not defined by just their survival. Rather, they are allowed to survive because they possessed the desired traits needed at the time where they were selected. Natural selection acts the same way as artificial selection. But, unlike events in the world of breeders, nature is governed by no purpose when it selects the "fittest." The traits of the fittest (plants or animals) are superior only at a particular environment in a particular time. Chances are that they would not be superior in a different environment. Biological evolution is best defined as adaptation of populations of organisms to an ever-changing environment. It offers no guarantee of general improvement. Yet, the Social Darwinists believe the struggle for existence is part of the struggle for continual change, for a kind of social progress they have confused with technological progress.

According to them, if a society is to progress, the basic laws of nature—which they misunderstand—must be obeyed. The elements of society "best adapted" would be those that would survive in open and unfettered competition. It follows that state interference in economics, state support of education, or state amelioration of poverty are to be avoided, for state control of any type tends to interfere with competition and so upset the process of natural selection.

The Social Darwinists first assumed that natural selection existed between individuals within a society, then they assumed that it also existed between social classes and nations. From this point, it was easy for them to take the next step, to accept the notion that there was natural selection between "human races." Inevitably they advanced the corollary that there were superior and inferior "races," the latter being destined to work for the profit of the former. The inferiority of certain races was an assumption that Social Darwinists thought could no more to be contested than the law of gravity, and they regarded it to be universal. But, as we have already seen, the idea was not original with   social Darwinists; they were expressing an old widely held notion. People throughout recorded history have tended to look upon themselves as "the real people" and all others as inferior to themselves, or even in some cases as not really human beings at all. All that the Social Darwinists did was to provide the idea with a pseudoscientific cloak.

Notes to Chapter Eight

1. Herbert Spencer, *Social Statistics*, 1862.

2. Darwin's theory of natural selection is based on three observations and two conclusions. First observation: Without environmental pressures, every species tends to multiply in geometric progression. A geometric progression is a sequence of numbers, each of which is obtained by multiplying the preceding number by a fixed number, called the ratio. For example, a population that doubles its number in a year has the ability to quadruple its original number in the following year and to increase eightfold in the third year and so on. (In this case, the geometric progression is 2, 4, 8, 16, 32 . . . .) Second observation: Aside from various fluctuations, the population remains markedly constant over a long period of time. First conclusion: Evidently, not all eggs are fertilized, not all fertilized eggs become adults, and not all the adults will survive and reproduce. Therefore, there must be some type of "struggle for existence." Third observation: Not all members of a species are alike. There is a great deal of individual variation. Second conclusion: Therefore, in this struggle for existence, those individuals whose variations are more favorable to survival naturally survive in proportionately greater numbers and produce offspring in proportionately greater numbers.

Chapter 9

ADAPTIVE OR NOT ADAPTIVE, THAT IS THE QUESTION

The question anthropologists of earlier generations never asked about what they called racial traits is a simple one: Are they adaptive or not adaptive? Not only have they assumed, and in some cases still assume,[1] that racial traits exist, but they concluded that each trait had an adaptive value. What do scientists mean when they use the phrase "adaptive value?" In order for a trait to have adaptive value, it has to improve the adjustment of the individual to the environment in which he or she lives; it has to be inheritable, and it has to enable its possessor to leave behind more fertile offspring than individuals who do not possess it. But in spite of the best efforts of numerous scientists over many generations, no racial trait has ever been shown to be adaptive, not even skin color, the most heavily used trait by those who engage in race classification.

For years, most observers assumed that dark-skinned individuals were favored by tropical climates and that light skinned individuals were favored by colder, less sunny climates, such as that of northwestern Europe. They were led to this assumption by the superficial observation that, in some parts of the world, but not in others, there appeared to be a small correlation between skin color and latitude, that is, the further a population is from the tropical region, the lighter its skin. From this general observation they were led to formulate hypotheses about how the so-called white race originated in western Europe and the black race in tropical Africa.

The questions posed by this line of reasoning are, in the last analysis, inappropriate and false because they presuppose the existence of races and conclude that races are adapted to the environments in which they live. A far more appropriate line of inquiry would be, first, to abandon these presuppositions. Second, we should investigate the mechanism by which our bodies provide skin color, which is the chemical melanin, This pigment, which we all have in our skin, is beneficial. If melanin is beneficial, however, is the possession of large or small amounts beneficial in specific and seasonal environments? This question is appropriate because it poses at least a partial answer that has nothing to do with race.

As to other racial traits, such as eye color, hair color, size of skull, form of hair, or shape of eyes, none has ever been shown to have given a comparative advantage to any individual or group. Most suggestions to the contrary have been speculative and often contradicted one another. This point was stressed by S. L. Washburn in his presidential address at the annual meeting of the American Anthropological Association, November 16, 1962:

> In the first place, in marked contrast to animals which are adapted to live in the arctic, large numbers of Mongoloids [Asians] are living in the hot, moist tropics. Altogether, unlike animal adaptation, then the people who are supposed to be adapted to the cold aren't living under cold conditions which are supposed to have produced them. They are presumed, as an arctic-adapted group following various laws, to have short extremities, flat noses, and to be stocky in build. They are, we might say, as stocky as the Scotch, as flat-nosed as the Norwegians, and as blond as the Eskimos. Actually there is no correlation, that is, none that has been well worked out, to support the notion that any of these racial groups is cold adapted.[2]

In this paragraph Washburn informs us there are animals that are adapted to cold weather but that no group of human beings is more cold adapted than any other. However, he does not suggest, or even hint, that there is something inherently wrong in trying to discover why some have characteristics that make them better adapted to hot or cold climates than others.

For example, camels are adapted to life in desert regions. Grizzly bears, on the other hand, certainly could not survive in such places. However, to say that Eskimos are cold adapted or that dark-skinned Africans are adapted to tropical climates, we are talking about human populations, not separate species of human beings. Africans can survive in Alaska, Eskimos in Africa. The differences in adaptation, if they exist at all, are minimal. One wonders if those who looked for adaptation of human groups in fact considered human races to be separate human species.

Notes for Chapter Nine

1. C. Loring Brace, one of the first anthropologists to deny the existence of human races, believed in the adaptive value of racial traits. See his paper, "A Non-Racial Approach Toward the Understanding of Human Diversity," in *The Concept of Race*, ed. Ashley Montagu (New York: The Free Press, 1964).

2. S. L. Washburn, "The Study of Race," in *The Concept of Race*, ed. Ashley Montagu (New York: The Free Press, 1964).

Chapter 10

WHY DIFFERENT SKIN COLORS?

It is yet to be demonstrated that the average Arab, for example, has too much skin pigmentation to thrive in Helsinki or too little for Dakar.

Richard Lewontin [1] (1982)

Why does human skin color vary from very dark to very light? A clue to the answer for this question lies in the observation that, before the great era of human migration that followed the European voyages of discovery, dark-skinned peoples, in general, were found in areas of high solar radiation; light skinned people, in general, lived in lands further north, lands with cool, cloudy climates. This observation suggested to many that in some way these peoples had the right skin pigmentation for the climatic conditions in which they lived.

The migration of people that has occurred during the past five centuries makes it difficult to undertake any geographical or ecological study of the distribution of skin color throughout the world. Nevertheless, this trait seems to have been associated with two climatic and environmental factors: First, darkness of skin color tends to be inversely related to latitude and temperature; the closer a population is to the equator and the higher the temperature, the darker the skin color tends to be. Second, the intensity of solar radiation is directly related to latitude and altitude; the closer the location to the equator and the higher the altitude, the greater the intensity of the radiation, in particular radiation within the ultraviolet range.

The distribution of deeply pigmented people in areas of intense ultraviolet radiation makes adaptive sense, since there is strong evidence that melanin in the skin protects against the harmful effects of ultraviolet radiation. It protects from sunburn and ultimately from cancer. Deposited in the top layer of the human epidermis, it prevents ultraviolet light from reaching DNA, the chemical of heredity, that exists in the live skin cells, which otherwise might be damaged by ultraviolet radiation.

Skin completely devoid of melanin, such as that of albinos, or skin with little pigment and incapable of tanning, such as that of many people with red hair, is far more susceptible to ultraviolet damage than pigmented skin. But, even the skin of highly pigmented people, when exposed to the sun for many years, is susceptible to skin cancer. In this process DNA in skin cells is altered, causing them to divide uncontrollably and form tumors. This damage first becomes visible as a small, scaly, precancerous spot, usually on middle-aged or older people and in areas of the skin generally not protected by clothing. In time these spots can turn malignant, becoming what are called translucent basal-cell nodules that slowly expand into adjoining tissue.

It seems, therefore, that the possession of melanin in the sunny parts of the world is a biological advantage. However, this does not necessarily mean that an increase in melanin in the skin of people in the tropics would naturally occur. We still have to show how natural selection would favor an increase in pigmentation under these conditions.

Though melanin protects against sunburn, except in a few cases sunburn is not a fatal disease but a temporary discomfort.[2] There is no evidence that people who are susceptible to sunburn have fewer children than those who do not sunburn. It is therefore doubtful that freedom from sunburn was a factor in increasing the amount of melanin in the skin.

Protection against skin cancer is another advantage of dark pigmentation. It is not clear, however, how strong natural selection from this source is. Sunlight-induced cancers are of low malignancy, and they occur mostly late in life, long after an individual has reproduced his or her genes for light-pigmented skin and susceptibility to cancer. As Richard Lewontin said,[3] ". . . even the most devoted 'white' sun worshipper tanning on the beaches of Southern California does not develop skin cancer until middle age (after having the opportunity to reproduce as many offspring that he or she wishes). If early hunting and gathering societies were at all like present-day ones, all children would have been produced well before the age at which skin cancer would develop. Nor can one invoke the loss of support of children resulting from the premature death of parents, inasmuch as the highly cooperative nature of primitive societies protects orphaned children."

Though it is true that skin cancer generally appears late in life, recent literature suggests that mortality from skin cancer sometimes does occur before the completion of the reproductive period. In Australia, a region of high solar radiation, deaths due to skin cancer among light skinned children have been recorded.[4] These deaths occurred despite medical care. So, one would expect that among populations without access to adequate medical care, the rate of mortality would be even greater. Hence, we cannot completely rule out the influence of natural selection for dark skin color in highly sunny climates.[5]

However, there are two ways to possess a dark skin in highly sunny climates. One is to be constitutively dark skinned, that is, to be born with dark skin. The other is to acquire a dark skin by getting a tan. As we have already emphasized, there are marked differences in the ability of fair-skinned individuals to tan. Some individuals tan well and fast, others tan little or not at all. A good tan is as effective as a constitutive dark pigmentation in preventing sunburn and protecting the skin from the ravages of cancer. Hence, any discussion of the adaptive value of skin color should include the subject of tanning ability. People like the Arab to whom Richard Lewontin referred in the epigraph of this chapter, who is light skinned part of the year and highly tanned the rest of the year, may have advantages not only over those who cannot tan but those who are constitutively dark skinned. What are the advantages, if any, of being light skinned in the winter time?

If it is clear that having a large amount of melanin in highly sunny climates is an advantage, we are not sure about the advantage of having little pigment for those in northern climates of Europe, Asia, or America. For many years, a widely held hypothesis among anthropologists was that fair skin among the inhabitants of northwestern Europe was a long term adaptation to cloudy skies and a cool climate with little ultraviolet radiation. A decrease in the amount of melanin permitted ultraviolet radiation to reach the inner skin and synthesize vitamin D, which is essential for proper calcium metabolism.

This vitamin controls absorption of calcium through the intestine, regulates excretion by the kidneys, and aids in deposition of minerals in our bones. These functions are so critical that the lack of vitamin D causes a bone defect: rickets in children and osteomalacia in adults. Unlike other vitamins, it is not present in significant amounts in the normal diet. It occurs in small amounts in vegetables and in large amounts in the liver of bony fishes. Before the incorporation of vitamin D into the milk supply, the main source for children was cod liver oil.

Most vitamin D received by the human body is synthesized by action of ultraviolet radiation on 7-dehydrocholesterol, a compound present in the lower layers of the epidermis. Since melanin prevents ultraviolet radiation from reaching these layers, a high concentration of the pigment is likely to reduce the skin's ability to manufacture vitamin D.

Hence, it would seem that decreased melanization would be advantageous for increasing the efficiency of vitamin D formation in northern latitudes where sunlight is limited during the winter months and where the body is generally clothed against the cold. In these climates, vitamin D synthesis would depend on the exposure of a small area of skin to relatively small amounts of ultraviolet light. Two facts tend to support the vitamin D hypothesis of skin color adaptation. First, in the nineteenth century in industrial cities of Europe where smoke obscured sunlight and prevented vitamin D formation, there was a prevalence of rickets among children. The argument is weak since rickets also could have been caused by an absence of calcium in the diet. Second, before the advent of vitamin D enriched milk, the rate of rickets in northern U.S. cities was more conspicuous among children with dark skin than among the light skinned children. The vitamin D hypothesis could also account for the fact that women, in general, have lighter skin than men. Given their need for more calcium during pregnancy, this makes good adaptive sense. Let us assume that having a fair skin is an advantage in northern climates because it maximizes the synthesis of vitamin D by ultraviolet light, which is at a premium during the winter months. We still have to show how natural selection would favor a decrease in pigmentation. We have referred already to rickets. It is characterized by bone defects in children that are caused by a lack of adequate vitamin D. Though its victims usually survived, distortion of the pelvis in growing girls causes them later to experience difficulty in natural childbirth. In the past, these difficulties could have resulted in death for the mother and the child. It has been calculated that under such conditions, if one child out of one hundred died because of vitamin D deficiency, it would have taken less than 50,000 years for selection favoring lighter skin[6] to take place.

The vitamin D hypothesis could also explain why the skin of children tends to be less dark than that of at least one of their parents and that it darkens throughout childhood.

The fact that many of the arctic peoples, who spend six months or more of the year in the dark, have quite dark skins was disconcerting to the proponents of the vitamin D hypothesis. They should have been vitamin D deficient, but they were not. To explain this, it was proposed that the arctic people get their vitamin D from their fish diet.

For a while, the vitamin D hypothesis seemed irrefutable, capable of overwhelming almost any objection. However, during the past two decades, evidence has been accumulated which brings into question the idea that vitamin D deficiency was responsible for the appearance of light skin in Europe. The evidence came from physiological and anthropological research. The arguments against the vitamin D hypothesis can be summarized in this way: If the hypothesis were correct, we would expect rickets to have been more prevalent in the past than it is today for two reasons. First, the European climate has been more rigorous than it is today and our ancestors of hundreds or thousands of years ago had no notion of how to combat the ravages of vitamin D deficiency. However, the disease seems to have been nonexistent in the past; in fact, no evidence of it has ever been found in European skeletal material. If the vitamin D hypothesis were correct, one would expect rickets to be prevalent. But, in fact, the disease is virtually absent from Europe's rural areas and was relatively common in cities only after the Industrial Revolution introduced smoke and pollution into the atmosphere.

Other arguments have been advanced against the vitamin D hypothesis.[7] At high latitudes much of the earth's surface is covered by snow, and snow is a potent reflector of ultraviolet radiation, making the environment a high-intensity field of UV radiation, which in turn makes it possible for any dark-skinned individual to get enough vitamin D for his or her survival. If the vitamin D hypothesis were the *real* explanation for the evolution of skin color in northern Europe, how would it explain the fact that British residents of West Indian origin rarely manifest clinical rickets though they have deeper skin pigmentation than British Asians, who are more susceptible to rickets and osteomalacia. In the United States, dark-skinned children are now believed to be no more prone to rickets than light skinned children. It has been found, however, that children who had lower concentrations of vitamin D were those who lived in overcrowded dwellings where they were confined indoors and deprived of sunlight. In other words, their rickets might well be due to socioeconomic factors rather than to the amount of melanin that blocks UV light.

However, the biggest blow struck against the vitamin D hypothesis is the discovery that our body can store vitamin D during the spring and summer months and utilize it during the winter months. This ability for vitamin D to be stored has been recently demonstrated by studies of light skinned British children who took summer seaside holidays, and of gardeners who worked throughout the year in Scotland, where daylight hours are short in the winter months. In both cases, the vitamin D levels during the months of November and December were far higher than those of the controls, that is, individuals who were indoors. Although the human subjects in these studies were light skinned, it is believed that these results are also valid for people with dark skin. While light skin would produce its maximum level of vitamin D in thirty minutes, dark skin would produce it in three hours. Over several months of summer, this time difference would not be significant. Even if our ancestors living in an exceptionally cold and dim Europe were dark skinned, they would have received enough UV light throughout the year. As to the Scottish gardeners, they would have continued to produce and store vitamin D as late as the fall months and use it in the winter.

On the whole, the theses that low melanin content helps ultraviolet light reach the skin to manufacture vitamin D, and that a high melanin content in the skin is an adaptation to life in the tropics because it offers protection against sunlight, have little to recommend them. It is true that individuals who have no melanin or little melanin in the skin suffer from ultraviolet injury. But the possession of a large amount of pigmentation does not seem to make very much difference one way or the other. Furthermore, the possession of melanin in tropical regions is not always beneficial. It has been shown that dark pigmented individuals do not get rid of the heat as well as light pigmented people do; in other words, they do not cool as easily as light pigmented do. In this respect dark skin would be more beneficial to a person living in Sweden than to someone living in Senegal: The former could use the extra calories to fight the cold, while the latter has to eliminate them to fight the excess heat.

If vitamin D were not responsible for the presence of light skin in western Europe, what other factors could have been? One may have been frost injury. It has been known since World War I that "black" soldiers were more prone to frost injury than "white" soldiers. This was clearly established by careful epidemiological studies during the Korean War. A few years ago, the idea that cold injury may have been a factor in the decreasing skin pigmentation among the inhabitants of northern Europe was tested.[8] Investigators took black-and-white piebald guinea pigs, anesthetized them, removed their hair by plucking, and then froze pigmented and nonpigmented skin samples from each animal. They did this in order to attribute any difference observed to the presence of melanin and not to individual variation among the animals. They found that pigmented cells were damaged more seriously than nonpigmented cells, indicating that melanin may be a significant factor in frost injury.

Of all the traits that differentiate us from one another, skin color has been the most studied. Yet, we have very little understanding of what the advantages are of having more or less melanin in our skin. If we know little about the adaptive significance of skin pigmentation, we know even less about that of other characteristics that distinguish us, for example, the shapes of noses, hairiness, eye and hair color, or the shapes of our eyes. Could it be that the reason we are unable to discover their adaptive significance is because they have none?

Notes for Chapter Ten

1. Richard Lewontin, *Human Diversity*, Scientific American Library (New York: W. H. Freeman, 1982), p. 127.

2. H.J. Blum, *Carcinogenesis by Ultraviolet Light* (Princeton, N. J.: Princeton University Press, 1959). Also "Does the Melanin Pigment of Human Skin Have Adaptive Value? An Essay in Human Ecology and the Evolution of Race." *Quarterly Review of Biology* 36:50-63. 1961.

3. Richard Lewontin, *Human Diversity*, p. 126

4. Frederich Urbach, John Eptsein, and Donald Forbes, "Ultraviolet Carcinogenesis: Experimental, Global and Genetic Aspects," in *Sunlight and Man*, ed. Tomas Fitzpatrick (Tokyo: University of Tokyo Press, 1974).

5. There have been two other hypotheses to explain heavy pigmentation in tropical climates. For a long time it was thought that too much vitamin D caused too much calcification of bones and cartilage and that a large amount of melanin in the skin prevented the formation of too much vitamin. However, no case of vitamin D toxicity due to prolonged exposure to the sun had ever been found. This was explained by M. J. Holick in 1981. He found that if the skin is exposed to sunlight for a long time, previtamin D-3, a precursor of vitamin D, is changed into two non-functional compounds that are eliminated. This hypothesis might have to be abandoned.

   A second hypothesis to explain why melanized skin has been selected in the tropics is highly plausible but has not been pursued. According to Branda and Eaton, "Skin Color and Nutrient Photolysis," *Science* 201 (18 Aug 1978): 625-6, a large amount of melanin may protect against photolysis of crucial light-sensitive vitamins and other products by ultraviolet radiation. To test their hypothesis, they used folate (a derivative of the vitamin folic acid) as a model. They found that exposure to ultraviolet light does decrease the amount of folate in the blood. Since deficiency of folate, which occurs in many marginally nourished populations, causes severe anemia and infertility, it is possible that prevention of ultraviolet photolysis of folate and other light-sensitive nutrients by a dark skin may be one reason for maintaining this characteristic in human groups indigenous to regions of intense solar radiation.

6. Frank B. Livingstone, "Polygenic Models for the Evolution of Human Skin Color Differences," *Human Biology* 41: 480-493 (1969).

7. Ashley Robbins, *Biological Perspectives on Human Pigmentation* (Cambridge: Cambridge University Press, 1991).

8. Peter Post, F. Daniels, and T. Binford. "Cold injury and the Evolution of 'White' Skin." *Human Biology* 47: 65-80 (1975).

Chapter 11

WHY DIFFERENT SHAPES OF NOSES? WHY SO MUCH OR SO LITTLE HAIR ON THE BODY OR THE HEAD?

My nose is huge, enormous, vast!
Listen, poor snub-nose, flat head, addle-pate,
Here is an accessory I'm proud to wear;
For a large nose betokens a large heart.
Symbol of courage and of courtesy,
It indicates a nature kind and keen,
Witty and warm and liberal—like mine
And never one like yours, you stupid oaf!
Because your foolish features are as bare
Of pride, of passion, and of purity,
Of inspiration, even of a nose—
As that on which I now will plant my boot!
            Cyrano de Bergerac
            Edmond Rostand[1]

One of my friends has a large, convex, highly bridged and curved nose. When he was a graduate student years ago, he visited a biochemical laboratory in Detroit. The man who was in charge of the laboratory was a Jew who seemed to have a stereotypic view of what Jews are supposed to look like, for at the end of the visit, pointing to his own nose, he asked one of his colleagues, also a Jew, if my friend was "one of us." Well, my friend was not "one of them," for there are many people throughout the world who have noses like his and are not Jewish. This type of nose is found among much of the Middle Eastern population, especially in the mountain regions lying between Israel and India; in Afghanistan, for example. But this type of nose is also found in parts of the world other than the Middle East: in the British Isles among the Scots, in North America among Indians, in the Bavarian mountains of Germany, and in Italy where, in a less accentuated form, it is called a Roman nose. Our noses—distinct features of our faces and personalities, if we believe Cyrano—are good examples of a structure in which bone and soft tissue combine to produce a highly varied and often irregular outline. The upper half of the nose is a long, bony protuberance covered by a uniform and rather thin layer of tissue. The lower half is composed of cartilage and some fatty tissue. The bony portion of the nose has two parts: the root, which is the portion lying between the eyes, and the bridge, which continues down the junction with the cartilaginous portion.

The extent to which the root of the nose projects forward is remarkably variable. The bridge has its own variations. It can simply continue the profile of the root or curve forward in a hump, drooping a little at the end. The bridge may be distinctly wider than the root. The lowest part of the nose includes the tip and the two wings that surround the nostrils. The shape of the nasal tip is roughly spherical and varies greatly in size. A small tip makes the nose appear pointed; a large tip makes it appear rounded or blunt in outline, and makes it look knobby from the front. The combination of root, bridge and tip variations produces a virtually infinite variety of shapes.

Europeans tend to have noses with prominent bridges and a cartilaginous part that is not quite equal in dimension, so that the general profile of the nose is convex. Asians tend to have noses with a low root and a bridge that continues the line of the root, and with a tip of modest size so that the whole nose is low lying and straight. Africans tend to have noses that have a moderately low root, which are short from top to bottom and relatively wide across the wings.

The above paragraph contains a description of nose types that populations tend to have. In fact, our noses vary a great deal in size and shape; no one shape or form is possessed typically by any single human group. Scientists, who always feel confident when they are able to measure things, have invented the nasal index (a ratio of the length to the width of the nose) to describe the shape of a nose. For example, an index of 104 would describe a nose slightly wider at the nostrils than it is long. Such an index is found not only among the pygmies of the Iteri forest region in central Africa but also among Australian aborigines. Narrower noses, represented by low indices (85 and below), are found among many groups throughout the world: American Indians, North Africans, Europeans, Eskimos, and Black Africans, in particular those from Ethiopia. That every shape of nose is found throughout the world suggests that nasal shape is not a racial trait, that is, not a trait which one human group specifically has.

There has been much speculation about the adaptive value of different forms of noses and whether noses have developed because of particular advantages in particular environments. It has often been said that a long and narrow nose is an advantage in cold climates because it increases the surface past which air flows as it is breathed in, warming and moistening it before it goes into the lungs. There are at least two problems with this hypothesis. First, our noses, long or short, contribute little to the total warming and moistening of the air that reaches our lungs. Second, and more important, there is no evidence that dry or cold air is harmful to the lungs. If the hypothesis were correct, we would expect that Australian aborigines, who live in dry climates, would have long narrow noses. But they do not. We would expect animals living in cold climates to have long snouts. But they do not. It has also been suggested that the low-profile nose found in central and eastern Asia was an adaptation to cold climates. According to proponents of this hypothesis, a nonprotruding nose would be an advantage, since any protuberance of the body is among the first parts of it to get cold because of the large amount of exposed surface and the difficulty of circulating enough warm blood through it to counter heat loss. Though some correlation has been shown to exist between nasal index and air temperature and humidity,[2] no evidence has been shown that low noses have an advantage over long and narrow noses in cold climates. It is not enough to have correlations between two variables; one has to do experiments to show that one variable is the cause of the other. After all, I can find a nice correlation between my aging in years and the price of new cars, but I doubt that anyone can show that either one is the cause of the other.

The only practical advantage that has ever been found for a specific shape of nose is that a convex nose has been found to be helpful in preventing eyeglasses from dropping further than they should. Everyone would agree, however, that this advantage has nothing to do with the adaptive values of nose shapes. That we do not find any correlation between nose shape and environment is not surprising. Every shape of nose can be found around the world and functions as well there as any other place. In other words, we are well adapted to any environment of the world regardless of the shape of our noses. The same is true in regards to the amount of hair we have on our bodies or on our heads.

Very early in life I became conscious that the amount of hair on the face and body varied considerably among individuals. As a young man, I became conscious that it varies as well among human groups. When I was a student, I had a roommate from Iran who was the hairiest man I ever met. His body was covered with hair and he was already bald. It took him a long time to wash himself. However, when he was in a bathing suit on the beach, all the girls could not prevent themselves from looking at him, though he was certainly not a handsome man. Then, a few years later, when I was a graduate student, I worked and sometimes roomed with a Chinese student who used to say, "It is Tuesday, I better shave today, whether I need it or not."

The very slight amount of hair on the face and on the body of my Chinese friend was characteristic of people from Asia. The very large amount of hair on the body of my Iranian friend was exceptional, though the people from the Middle East have strong beards and have, in general, more hair on their body than Asians or Africans, but far less than the Ainus of northern Japan or the Australian aborigines. Why does the amount of body hair vary among peoples of the world? Perhaps it would be an advantage for populations living in cold climates to have hair on their bodies. If this idea had any merit, we should find populations in northern climates with the greatest development of body hair. But this is not the case. Eskimos and peoples of Siberia and of Tierra del Fuego have little body hair.

On the other hand, it was suggested more than forty years ago that a lack of hair on the face was an advantage. In extreme cold, a beard may be encrusted with frozen vapor from the breath and thereby become a nuisance.[3] However, this hypothesis is far fetched in the light that brown bears, with their long and heavy coats, are not hampered in their search for their favorite food in the freezing rivers of Alaska. As a matter of fact, there is no evidence whatsoever that men with heavy beards are less prolific in very cold climate than those who are beardless. We have to conclude that length, presence or absence of hair in human beings has no adaptive value. These differences among people seem to be simply different expressions of a vestigial trait whose function in our ancestors of long ago was to protect them against the cold, but which in us has no function.

As I mentioned, my Iranian friend was bald. This seems to be a characteristic of the people of the Middle East. Balding is also common in Europe, but is rare among Asians, American Indians and blacks from Africa. Again we have no idea why balding is more prevalent in some parts of the world. It is hard to believe that it might have some adaptive value. On the contrary, as the author is well aware from his own experience, it has a lot of disadvantages. One of them is that the unprotected top of the head of bald people is often hurt when they hit a low ceiling, or mow their lawn under a low branch of a tree, or, when working on their car engine, the hood accidentally drops on them.

There are many jokes and sarcastic comments made about baldness, but there are also myths that surround it. One of them is that bald men are more sexy than nonbald men. This myth might have originated with Aristotle, who thought that sexual intercourse caused baldness. But Aristotle was wrong. Scientists have shown that one form of baldness, pattern baldness, is due to a single dominant gene, but the male hormone testosterone is necessary for its expression. However, the amount of this hormone produced in bald men is not greater than that produced by men who are not bald. In this sense, they are no more sexy than other men, and the number of children they have has nothing to do with the fact that they are bald. Baldness, like shape of nose, does not seem to have any particular adaptive value.

What we have just said about nose shape and distribution of hair on our body and face is true for other traits such as eye and hair color, traits that we all talk about and know little about.

Notes for Chapter Eleven

1. *Cyrano de Bergerac.* A heroic comedy in five acts by Edmond Rostand. English version by Louis Untermeyer. (New York: The Heritage Press, 1954).

2. J. S. Weiner, "Nose Shape and Climate," in *Readings on Race*, ed. Stanley Garn (Springfield, Ill.: Charles C. Thomas, 1968).

3. G. S. Coon, S. M. Garn, and J. B. Birdsell, *Races: A Study of the Problems of Race Formation in Man* (Springfield. Illinois: Charles C. Thomas, 1950).

Chapter 12

WHY DIFFERENT COLORS OF EYE AND HAIR?

"I knew she was not her daughter, for her parents had blue eyes and she has brown eyes."

That is what the priest-detective Father Dowling said to the resourceful Sister Steve in a recent episode of the television series *The Father Dowling Mysteries*. Though Father Dowling's deduction led him to find the murderer, it was not scientifically sound. The inheritance of eye color is far more complex than is generally assumed. However, we cannot blame Father Dowling or his script writers for repeating common folk wisdom, which has been taught for years, even in beginning courses on genetics. Eye-color inheritance was considered to be a classic example of a human trait determined by one pair of genes; one gene for the brown color was assumed to be dominant over the gene for blue. Not only was it taught that way, but students were given problems to solve which were based on a simple, but incorrect hypothesis. The hypothesis is contradicted by the fact that there are many documented cases of children with brown eyes whose biological parents have blue eyes. It is indeed an oversimplification leading to misinterpretations of family histories, such as the one that concerned Father Dowling. Sometimes the suspicions may have been justified. But in the great majority of such cases, doubts of parenthood were groundless.

That the simple explanation of eye color inheritance is incorrect can be demonstrated very simply by realizing that eye color comes not only in blue or brown but in many other intermediate shades—hazel, green, and gray as well. The color of the eye is due to a number of factors, but variation in them appears to be largely due to the amount and distribution of melanin, the same pigment that gives skin its color. When we talk about variations in eye color, we are restricting our discussion to one part of the eye, the iris. The iris is a round, contractile membrane which controls the amount of light reaching the retina in the back of the eye. Its color is the result of the production of melanin in its tissue, in particular, in the front of the iris. Except for albinos, who have no pigment in their eyes, we all have some. In those of us who have blue eyes, no melanin can be seen in the front of the iris, but it is present behind it; in other words, there is no "blue" pigment. The eyes appear blue because of the same optical effect that causes the sky to appear blue on a clear day. Dust particles in the atmosphere scatter light with short wavelengths, such as blue light, but not light with longer wavelengths, such as red. In blue eyes, the minute protein particles in the iris play the same role as dust particles in the atmosphere. As light traverses the relatively melanin-free front layers of the iris, they scatter the short, blue wavelengths to the surface. In this regard, poets are entirely correct when they liken blue eyes to the blue of the sky.

If the front of the iris has a little pigment, the color of the eye will appear blue flecked with brown. If the pigmentation is thinly dispersed, it might form a yellow filter which, with the blue optical effect, would give a greenish tint. A little more pigment combined with the blue optical effect would give a combined shade that is called hazel. With progressively larger amounts of pigment in the front of the iris, hazel grades into dark brown or even very dark brown (black).

Obviously, eye color is complex. Not only does it depend on the amount of melanin present but also on the color and brightness of the light striking the eye. In many cases, whether we say a person is blue eyed or brown-eyed depends on how closely we examine his or her eyes and under what conditions of illumination. Eye color also varies with age and sex of the individual. Generally eyes darken, from birth to maturity, and thereafter begin to lighten with aging. Women tend to have darker eyes than men.

Most human populations are dark-eyed. Although, as a result of mutations, blue or gray eyes do occasionally occur in every population, light eyes are most common among the peoples of Europe and their descendants. This has suggested to anthropologists that blue eyes conferred some sort of advantage under local conditions. But none has ever been found.[1] One thing is clear, however: people who have light-colored eyes see clearly but complain of acute discomfort when confronted with very bright illumination. Today, these people are the main support of the sunglass industry.

On the other hand, it was first believed that individuals having dark eyes had greater visual acuity in bright light and hence would be better adapted to stressful environments such as deserts and snow fields. But this hypothesis was crushed when it was found that the density of eye pigmentation does not seem to influence visual acuity under increasing conditions of brightness.[2] Individuals do differ in visual acuity, but this depends on factors other than eye color.

However, proponents of the idea that eye color has adaptive value did not give up. They turned their attention to animals. Cats, both small and large, have fairly light eyes. They hunt effectively both in full daylight and at very low levels of illumination. The eyes of northern wolves and arctic breeds of dogs are light colored, either gray or blue. So are the eyes of a species of macaque, the snow monkey of northern Japan, which has blue eyes as an infant and yellow eyes as an adult. This is in contrast to its many close relatives farther south, who have brown eyes. Although this suggests that a small amount of pigmentation may have some advantage in parts of the world where light is low, this hypothesis has never been supported by any experimental evidence.

We should note that in order to explain the distribution of light-eyed people in northern Europe and dark-eyed people in the rest of the world, we would have to demonstrate not only the importance of eye pigmentation under certain climates, but also to show that people having a specific pigmentation leave more offspring under certain climatic conditions than under others. There is no evidence that this is the case.

What I have just said about eye color is also true of hair color. Like eye color, hair color is very complex, as indicated by the great variety of natural colors. It varies with the amount of pigment present, light intensity and color, and the age of the individual. As in the skin and the eye, the pigment in the hair is melanin. Dark hair, like dark eyes, predominates in the world, and it is mainly in the northwestern part of Europe that the incidence of blond and red shades is high. Red hair is said to occur in about 3 percent of the population of England and in 11 percent of the population of Scotland. However, even in these regions light hair is by no means the rule in adults, for the surveys of hair color were made among schoolchildren and it is well known that hair color becomes more intense from childhood to adulthood. This is striking among Australian aborigines. They generally have dark brown hair as adults, but occasionally have children who are born fair-haired and who remain so until they approach puberty, when their hair darkens as a result of an increase in the amount of sex hormones. This pattern is also common among Europeans.

The fact that dark hair predominates under all climatic conditions leads us to believe that hair coloration has no adaptive value. Yet, it has been suggested that dark hair is useful in intercepting solar radiation that might damage the deeper layers of the scalp. However here also, no evidence has ever been found which shows that blond hair is not as efficient as black hair in this respect. We seem to be at a loss to find any advantage in having a specific coloration in our hair. Furthermore, we should not forget that for a characteristic to have an adaptive value, it has to permit its possessor to leave a larger number of children than the individual who does not possess the characteristic. According to Hollywood, blondes have more fun, but no one has shown that they have more offspring than brunettes or vice versa. Hence, it is doubtful that hair color has some adaptive value.

Since the color of skin, eye, and hair depends on the amount of melanin present in them, it is astonishing that there is a poor correlation between skin color, eye color, and hair color. After all, we would expect that the same factors responsible for the production of melanin in one body structure would show similar effects in another. Accordingly, we would have at the most, only three types of individuals—some with dark hair, black eyes and skin, others with blue eyes, blond hair, and light skins, and still others with intermediate shades. Unfortunately for "race" classifiers, this obviously does not occur. However, the exact opposite situation does not occur either. If the factors for determining pigmentation of hair, skin, and eyes were acting independently of one another, we would expect that an individual with black skin would be as likely to have blue eyes and fair hair as an individual with fair skin. However, this is not what we find. What we have is something between the two extremes. Hence, in order to explain the partial correlation that exists between the various colors of skin, hair and eyes, we have to assume that there are some factors which act independently and others which act in concert, with the result that humans present an extreme diversity of pigmentation defying any classification.

We ended chapter ten asking the question, Could it be that we are unable to discover the adaptive significance of "racial" characteristics because they have none and that diversity is just that, diversity? At the end of this chapter, the answer to that question is obvious. Our physical differences in pigmentation, shape of nose, shape of teeth, amount of body hair, and so on, do not seem to have been important in our past and certainly are not important in today's world. Regardless of physical differences, any one of us can swim, ride a bicycle, drive a car, read or write, or work with a computer.

Notes for Chapter Twelve

1. *Journal of Ophthalmology* 63, 795-803.

2. See the symposium, "Anthropological Aspects of Pigmentation," April 11-12, 1974, published in the *American Journal of Physical Anthropology*, Vol. 43.

Chapter 13

RACE: GENETICISTS LED ASTRAY

Unfortunately, scientists failed to discard the old concepts when interpreting the new observations, or to use biblical metaphor, they put new wine into the old casks.

Albert Jacquard [1]

We have seen that during the eighteenth, nineteenth, and early twentieth centuries, there were many attempts to classify humanity into races by using skin color, hair type, nose configuration, skull shape, and many other physical traits. These attempts failed because scientists have never been able to agree on how many races there were or what a human race actually was. No genuine racial boundaries could be identified because there is tremendous variability of traits among individuals within any group that is established. However, in spite of repeated failures, most scientists remained convinced of the existence of human races. They did not abandon the idea of classifying mankind and hoped eventually to find a way to do this. Would it be the new science of Genetics?

Genetics teaches us that parents' characteristics are passed along to offspring in the form of units called genes. These genes, in conjunction with the environment, regulate the growth and the development process of each individual. In a few years, genetics had branched into diverse fields. One of these was *population genetics*. This is the field of biology that studies the genetic composition of plant, animal or human populations. The populations themselves, rather than the individual organisms, become the basic units of biological study.

In the late 1940s, anthropologists began to turn to population genetics, hoping that this new science would help them in their attempts to classify mankind into races. However, they failed once again and for the same reasons that they had failed earlier. Ironically, population genetics was very helpful in explaining how races in general form in plants and animals. The results of using this science should have led scientists to realize that there were no human races because the conditions needed for race formation never existed in humans. But for years scientists were incapable of seeing the implications of their own results. The reason for this was that they still had the same *idée fixe* as had their precursors, that human races really existed. They continued to ask that science confirm their preconceived idea and not whether the idea was correct or not.

Let us see how population geneticists and anthropologists went astray. The first geneticists thought that each trait was determined by only one pair of genes, one gene coming from each parent. This one gene, one trait hypothesis served a very useful purpose in promoting the understanding of the elementary cases of heredity. Soon, however, it was found that this hypothesis was inadequate because most human traits show a tremendous range of variability that could not be accounted for by just one pair of genes. Among the traits whose inheritance was complex were those that anthropologists had used to classify people. Further, it was shown that many of these traits were affected by a wide variety of conditions within the environment.

In order to be able to classify mankind into races, traits whose inheritance was simple and not affected by the environment had to be used. Hope was high that blood groups that had been discovered after 1900 would fulfill these requirements. Blood consists of a number of things, the most important of which, for our discussion, are the red blood cells that transport oxygen, and serum, a yellow fluid containing antibodies that defend the body against diseases. At the beginning of the twentieth century, it was discovered that some people's red blood cells carried one kind of substance, A; others carried another substance, B; some carried both substances; and in many individuals the red cells carried neither. Substances A and B were to be called antigens. Since then, more than sixty other types of antigens have been discovered. The function of these antigens is not known, but their genetics are simple and well known. There seems to be a direct one-to-one relationship between the blood antigens and the genes that determine them. This was only one of the reasons why anthropologists turned to blood groups in their quest for a better way of classifying humanity into races. There were four other reasons. First, blood groups do not seem to be influenced by the environment. They are genetically determined at conception or soon after and remain fixed for life. Second, blood types can be sharply differentiated from one another by a simple objective test. Third, millions of blood group determinations are made every year for medical reasons, thus providing a wealth of free information about blood group distribution in humans all around the world. Fourth, for each blood group, human populations often differ in their genetic makeup.

How, one might ask, is it possible to describe the genetic makeup of an entire population? It is not quite as difficult as it may seem. Let us assume, for simplicity's sake, that we are studying a specific blood group called MN in a population of one hundred human beings. Each individual has two genes for the MN blood type: one that he or she inherited from one parent, the other from the other parent. These genes may be identical or different. Different forms of a gene are called *alleles*. In our example we have two alleles, M and N. One is responsible for a specific antigen, the other for a different one. There are only three possible types of individuals in this blood group. Some individuals in our population are of type MM, others are NN, and still others are MN. Such a population is genetically characterized by the number of M alleles and the number of N alleles. The total number of genes for the MN blood group in our population of one hundred human beings is two hundred. Geneticists call this collection of genes the *gene pool* of that population. The number and kind of each allele found in the population determine the unique characteristics of the gene pool.

Let us assume that in our population there are sixty MM individuals, thirty NN individuals, and ten MN individuals. The proportion of the M alleles in the gene pool is 120 (each of the sixty MM individuals furnishes two M alleles to the pool) plus ten (each of the ten MN individuals furnishes one M allele), for a total of 130 in 200, or 65 in 100. In the same manner we can calculate the proportion of the N alleles in this population: It is 35 in 100. Geneticists prefer to use frequencies instead of proportions. In this case, the *gene frequency* for the M allele is 0.65 and that for the N allele is 0.35. Gene frequency is the proportion of a particular allele relative to all the alleles of a certain gene.

The first efforts at race classification based on blood groups were done by William Boyd,[2] who used the distribution of gene frequency of the three blood groups ABO, MN, and Rh. He obtained a classification of races similar to those based on considerations of geography and classical anthropology, adding some detail but not suggesting the reasons why "races" originated. Boyd had hoped that one population would have a particular form of a gene (and preferably more than one) that the others would not have. But this was not the case.[3] What he found was that each form of a gene present in one population was also present in all the other populations—only its frequency differed from one group to another. In fact, all Boyd had accomplished was to define races as populations differing in the frequency of some alleles of certain genes. His definition was a loose one and led to much speculation about how different two human populations would have to be before they might be considered races.

This was one problem that confronted population geneticists in their quest to classify mankind into races. Another originated from the discovery that populations may be similar in their frequencies of one gene, making them closely related, but dissimilar in their frequencies of another, making them distantly related. They tried to overcome this difficulty by using various statistical manipulations. But, without going into details, lots of information obtained this way has been largely contradictory.

Another fundamental problem originated when it was learned that one of the main assumptions on which blood group race classification was based was invalid. You recall that geneticists had assumed that the distribution of blood groups did not change when the environment changed. Unfortunately, when more data became available, this turned out not to be true. We know now that epidemic diseases can affect the distribution of blood groups. For example, smallpox, which in many epidemics has wiped out millions of people, seems to affect more persons of blood type A than persons of blood type B, with the result that the frequencies of the A allele will decrease while those of the B allele will increase. As another example, individuals of blood group O are apparently more likely than others to contract the Asian form of influenza and die from its effects. As a result of these epidemics, a population might appear to belong to a group different from that to which it could have belonged if the epidemics had not occurred.

Other studies have demonstrated that mosquitoes have preferences in the blood types they choose to feed on. People with type O blood are more often bitten than those with either type A or B blood, and thus have a better chance of getting malaria or other mosquito-borne diseases than others and dying. As a result, in countries infested with malaria, the frequency of blood type O would be smaller than in other parts of the world. Hence, the differences in gene frequencies in blood types now present around the world could be the result of selective effects of infectious or parasitic diseases that had affected mankind in the past. If this was the case, natural selection carried out by diseases would interfere with the process of unraveling "racial" history when using blood groups.

Population geneticists also discovered that the natural selection process is not the only factor that can change the frequency of a gene in a population. Other factors are migration, the size of the population, mutation, or a combination of all these. If we know only that the frequencies of the blood type alleles in two populations are different, there is no way to be sure which factor or combination of factors is responsible for the differences[4] and therefore we cannot infer the assumed "race" membership.

For all of these reasons, population geneticists failed in their attempts at race classification in humans. They failed for the same fundamental reason the earlier anthropologists failed: It is impossible to classify humans into more or less homogenous groups because human populations were never isolated from one another long enough to allow a significant genetic differentiation to take place. We cannot blame the population geneticists for this, but we can blame them for not realizing sooner that their own science explained why race classification in humans could not be done and and why the existence of human races is a myth.

In order to demonstrate this, let us go back to Boyd's definition of race. According to him, races are populations which differ in the frequency of some alleles of certain genes. His definition was adopted by population geneticists without realizing that their definition of race required them to abandon the traditional anthropological classifications. As Albert Jacquard wrote in the epigraph of this chapter, they "failed to discard the old concepts when interpreting the new observations, or to use a biblical metaphor, they put new wine into the old casks." To illustrate such a failure on their part, let us take the following example.

Imagine two populations, one in Detroit, Michigan, the other in Oakland, California. Both populations are "white." Let us assume that these two populations differ in the frequency of the gene responsible for the blood antigen A and that the difference is 0.8. Would population geneticists conclude that the two populations are races? They should, since the difference in gene frequencies between the two populations is large. But it is unlikely that they would, because these findings would go against their ingrained perception that all "whites" are of the same race.

Now, let us assume that the population in Detroit is "black" and the one in Oakland is "white." Let us assume that the difference in gene frequencies of the gene responsible for the blood antigen A between the two populations is 0.1. Would population geneticists conclude from these data that the two populations belong to the same race? They should, since the difference in gene frequencies between these two populations is small. But it is unlikely that they would, because these findings would go against their ingrained perception that "blacks" and "whites" are two different races. This is obvious from what well-known English population geneticist W. F. Bodmer wrote:[5]

> How much difference [in gene frequencies] does there have to be between populations before we call them races? After all, even the people of, say, Lancashire and Yorkshire are likely to differ significantly in the frequency of at least some polymorphisms [genetic markers], but we should hardly refer to them as different

races. On the other hand, most people would agree [but not the author of this book] that the differences between the indigenous peoples of the major continents, such as the differences between Africans, Orientals, and Caucasians are obvious enough to merit the label races.

One wonders why population geneticists bothered to calculate gene frequencies of populations to determine races if the old classification based on direct observation of physical differences was more influential than their gene frequency calculations. The answer to this question is obvious. Population geneticists, like the craniologists of yesteryear, were so sure human races existed [6] that they ignored their own data which contradicted their preconceived ideas. (The process of race formation will be clarified in chapter 17.)

Today, biologists have finally seen the light. They see races as subgroups of species. Race formation is a step toward species formation. Although the splitting of one population into two smaller populations that become two species is a gradual process, it is impossible to tell at a glance when populations become races and when races cease to be races and become species. We can, nevertheless, distinguish three stages:

1).   Race formation begins when the frequency of certain alleles of a gene become slightly different in one part of the population from what it is in other parts.

2).   If the differentiation is allowed to proceed unimpeded, most or all of the individuals of one population may, due to the process of mutation and selection, come to possess certain forms of genes which members of the other population do not. We are now dealing with distinct races.

3).   Finally, biological changes preventing the interbreeding of races may develop, splitting what used to be a single collective group into two or more separate ones. When such mechanisms have developed and the prevention of interbreeding is more or less complete, then we are dealing with separate species.

Since the differences between groups of humans are very slight, race formation in humans has never gone further than the first stage. The types of isolation mentioned do not occur and there has never been any complete social isolation, either. Migration and genetic exchange have always taken place between human groups. That is why human populations differ only in the frequencies of certain alleles, and this is not enough differentiation for them to qualify as being races.

For example, populations in central Africa have a much higher frequency of alleles that produce dark skin than do the European populations.

There are, however, people with dark skin in Europe and people with light skin in central Africa. The frequency of alleles for light-color eyes progressively decreases southward from Scandinavia throughout central Europe to the Mediterranean region and Africa. Nonetheless, some blue-eyed individuals occur in central Africa. The differences are relative, not absolute.

The reason for this is, as we pointed out previously, that mankind has never been split into isolated groups in the past, or if it was, as in the case of the Native Australians and Americans, there was not sufficient length of time for race formation to occur. Isolation, geographically or socially, is not likely to take place in the future. Groups from every continent have invaded every environment on earth. And with improvements in transportation and communication, we are more apt to break down more physical barriers between human groups than to set them up.

In conclusion, for hundreds of years scientists have tried to classify humanity into races, but have failed because—and it is worth repeating—our diversity is far greater on an individual basis than on a group basis. Although scientists assumed the existence of human races, they could not agree on what a race is. They were not able to list the criteria to be used in race classification, and since they could not, they could not agree on how many races there were.

Today, efforts to classify humanity have for the most part ceased. Scientists have finally realized that they were no more successful using blood groups, or other genetic markers, than they had been in the past when they were using skulls or skin color. They have finally realized that categorizing human beings into "races" requires such a distortion of the facts that its usefulness as a tool disappears. Simultaneously, the term "race" is disappearing from scientific writing because scientists no longer accept the clear cut division of humanity into white, black, yellow and red that is still present in most college curricula and textbooks. From a biological viewpoint, human races do not exist. This is a conclusion that most anthropologists and geneticists have accepted. Now they understand their task differently: to study human variability without the concept of race. It is also time for the rest of us to abandon this obsolete, destructive and false notion of race.

Notes for Chapter 13

1. Albert Jacquard, *In Praise of Difference: Genetics and Human Affairs* (New York: Columbia University Press, 1984).

2. William Boyd, *Genetics and the Races of Man* (Boston: Little, Brown and Co., 1950).

3. The gene for the blood antigen Diego has been found exclusively among Asians and American Indians of South America. Since it is rare even there, we are not sure that, if further studies were made, one would not find it in other populations of the world.

4. Though we have not been able to unravel the past history of "racial" groups, we have used blood frequencies to determine how much hybrid ancestry exists between two populations living side by side. Very exact estimates of the degree of intermixture have been obtained for "white" and "black" populations in the United States. We have also ruled out any extensive incorporation of American Indian alleles into the American black population. We have also confirmed that gypsies are isolates within the countries in which they live. They still retain the blood group distribution of their ancestors in India.

5. W. F. Bodmer, "Race and IQ: The Genetic Background," in *Race and Intelligence. The Fallacies Behind the Race-IQ Controversy*, Ken Richardson and David Spears, eds. (Baltimore, Md.: Penguin Books, 1972).

6. Geneticists were as reluctant as anthropologists to abandon the concept of race. For example, Theodosius Dobzhansky, one of the most famous population geneticists, said, "Some authors have talked themselves into denying that the human species has any races at all! . . . Just as zoologists are confronted with a diversity of human beings . . . Race is the subject of scientific study and analysis simply because it is a fact of nature." On the other hand, for years Ashley Montagu, single-handedly fought the battle against the dogma that race was a fact of nature. He was one of the first anthropologists to realize that the concept of race in humans was artificial, that it did not agree with the facts, and that it led to confusion and the perpetuation of error.

Chapter 14

RACE AND IQ:
A PSEUDOPROBLEM

"Race" and "IQ" are terms which seemingly possess a clear and well-defined meaning for millions of people. Their common usage implies the belief in a reality which is beyond question... Nevertheless, the truth is that these terms not only are unsound but in fact correspond to no verifiable reality, and have, indeed, been made the basis for social and political action of the most heinous kinds.

Ashley Montagu [1]

A wrong assumption that leads science astray may only delay its progress. It may have no other consequence. This was the case of the phlogiston theory in chemistry or the case of the theory of spontaneous generation in biology. However, the assumption that human races exist not only led science astray for quite a while but also had very harmful, even destructive social consequences. For years many scientists were convinced that different human groups differed in intelligence and they attempted to demonstrate that this was true.

What happened as a result of their efforts was not something in which we can take great pride. We have seen in chapter three that Morton finagled his skull data by convenient omissions and that Broca used "correction factors" when it was convenient for him to do so, but did not use them if they led to results that contradicted his preconceived ideas. With time, honesty prevailed and it was concluded that there were no skull differences among "human groups." Craniology collapsed. However, those who wanted to prove that there were differences in intelligence between sexes, human groups, and social classes did not give up. In the twentieth century they turned to IQ testing. Here, again, honest science gave way to bad science, or even to pseudoscience. It was bad science because those who believed that there were genetic differences in intelligence among human groups, "races" or classes were shown by their critics that they had not understood the nature of IQ, and that they had not grasped the fundamentals of genetics and of some of the mathematical methods they used to buttress their ideas. It was dishonest science because some proponents of the idea that there were differences in intelligence among human groups ignored the data that contradicted their views,[2] or even invented data to support their views.[3] Unfortunately, these are not just problems of the past; they are still with us today.

To understand these criticisms, let us first examine the nature of IQ tests. IQ tests have their origin in tests devised in 1908 by Alfred Binet in response to a request from the French Ministry of Education.[4] The ministry asked that Binet construct tests which would identify children whose lack of success in classrooms suggested the need for some form of special education. To do this, he devised a series of short, practical and diverse tasks that were related to everyday problems.

These tasks were to be taken in order of difficulty. Binet realized that a child had to reach a minimum age before he or she could perform some of his specific tasks. A child began with the simplest tasks and would proceed to the more complex, until he or she was unable to complete the next task. Binet called the age associated with the last task the child could perform his or her "mental age." By subtracting the child's mental age from his or her chronological age, Binet obtained a measure of the general ability of the child. Thus, he could identify the children who needed special education. They were those whose mental age was definitely behind their chronological age.

However, it soon became apparent that subtracting the mental age from the chronological age was not an appropriate mechanism for measurement. For example, a three-year difference between a mental age of two and a chronological age of five indicated a greater difference in development than a three-year difference between mental age of thirteen and a chronological age of sixteen. It would be far better, said the German psychologist Wilhelm Stern, to divide a mental age by chronological age. This quotient, he argued, would give the relative value of disparity between mental age and chronological age. Stern's idea was adopted with a slight modification in order. The quotient was to be multiplied by one hundred to get rid of the decimals. The number obtained in this way became known as the intelligent quotient, or IQ, of the individual.

Binet warned that his tests did not measure intelligence but helped to predict academic achievement; that they should not be used to rank children who were developing at about the same rate as others of their age; and that low scores were not to be used to stamp children as innately incapable, but as a warning that they needed help and special training. He further warned that tests developed for use in one culture could not be used in another culture.

All of Binet's warnings were disregarded by a large group of enthusiastic followers, especially those in America. One such enthusiast was H. H. Goddard, the scientist who introduced and promoted the idea of intelligence testing in the U.S. in the early twentieth century. He believed that the Binet tests were real indicators of intelligence, and interpreted test scores as relatively pure measures of innate intellectual potential. Goddard was very influential and, unfortunately, many of his ideas still persist today. Many people, not just ordinary citizens, but educators, social workers, and even a few psychologists, still firmly believe in the following three myths about IQ:

1.) Our IQ tells us how smart we are.
2.) Our IQ does not change throughout our life.
3.) Our IQ is totally inherited.

The view that IQ is the number which tells how smart we are can be found even in popular literature. For example, in the first page of the novel, *Forrest Gump*, we read: "I was born an idiot. My IQ is near seventy, which qualifies me, so they say."[5] This statement by Gump implies two things: First, that IQ is synonymous with intelligence, which is considered to be a single ability; and second, that this number will not change throughout one's lifetime. We also find these two assumptions made unconsciously in the commonly asked question, "What is your IQ?" Neither of these assumptions is correct.

IQ is simply a single score that represents the number of correct answers to a set of items which have been written to test how far some of our specific mental abilities have developed. Note that I did not say intelligence but specific mental abilities. This is because intelligence is not a substance secreted by the brain in the same way that bile is secreted by the liver—we cannot weigh it. It is not an attribute like height or weight that we can measure with a scale. In fact, intelligence is a very complex quality that is very hard to define. Psychologists disagree on what intelligence is. Some, a small minority, believe that it is a single broad ability. Others believe that it is a composite of many abilities. Binet shared this last view.

Accordingly, Binet developed quite a variety of tests, which were the basis of his scale. Even so, we know now that only a few of the intellectual abilities were represented by Binet's scale and the other scales that followed it. One of the abilities that "intelligence tests" generally miss is creative thinking. Among the abilities that modern IQ tests attempt to measure are those to learn, to understand, to carry on abstract thinking, to memorize, to carry out mathematical operations, and to master languages. But we also have other qualities, such as intuition, creativity, and common sense, which are much harder to test. Furthermore, the answers to the items on an IQ test largely depend on the knowledge the person has acquired through his or her social and cultural experiences, including the one of taking tests, on motivation, and on all kinds of other factors, known and unknown. Hence, an IQ score simply expresses one's performance on a particular test at a particular time. Such a performance can vary with an increase in knowledge and experience or any change in opportunities.

The fact that *Forrest Gump* scored seventy on an IQ test when he was a little boy might mean very little concerning his intelligence later in life. In fact, there have been many true cases of people who scored low on IQ tests when they were children but scored high as adults.

Because IQ tests measure performances which are influenced by many sociocultural variables, is it fair to give the same tests to people of different backgrounds? Obviously not. Many IQ tests give significantly different scores for children of varying cultural and socioeconomic backgrounds.[6] There were indeed a great many cultural biases in the famous Stanford Binet test, for it was devised with "white American middle class" children in mind, and we would expect that a child from a different background would answer the items in the test very differently. In fact, in many cases, we would expect an intelligent child from certain backgrounds to give a particular "wrong" answer. On the other hand, it is possible to devise a test that middle class children will fail. R. L. Williams,[7] a "black" psychologist, has illustrated the point by developing an IQ test that asks questions about things familiar to any ghetto child, but unfamiliar to the middle class white child.

To overcome this limitation, writers of IQ tests have attempted to devise a culture-free test in which the students' answers do not depend in any way on the environment in which they grew up and developed. But test developers failed. The testing of children reared in different cultures or subcultures presents many technical difficulties. Each culture fosters the development of a different pattern of abilities. Tests constructed within a particular culture reflect such an ability pattern and thus tend to favor individuals reared in that culture. For example, the mere use of paper and pencil or the presentation of abstract tasks that have no immediate practical significance will favor some cultural groups. Other relevant factors that differ among cultures or subcultures include extent of familiarity with pictures, drive to excel, or simply motivation to take the test.

The third myth about IQ, that it is in great part inherited, may be the most fallacious. We can talk about inheritance of human traits, such as height or eye color, though technically what we actually inherit are the genes for these traits. But to talk about inheritance of

IQ, as some psychologists and even geneticists do, seems rather strange.

An IQ score, as we have already emphasized, reflects the result of a performance on a specific test at a particular time. It is a number, the result of an attempt to measure intelligence. It is not a human characteristic. If one cannot talk about inheritance of IQ, can one talk about inheritance of intelligence? The answer is yes, for certainly some part of our capacity to learn, to grow and change is inherited, as all of our human characteristics are. However, we have no way to isolate such a part, which has been loosely called "innate intelligence."

The main reason why IQ tests, or any other tests, cannot measure "innate intelligence" is that one is not born with the ability to think. Such an ability develops only by exposure of the brain and nervous system to proper conditions: good nutrition before and after birth, a family environment that stimulates all of the senses of the child and supports and encourages that child before he or she goes to school. Since we are not born with the ability to think, it makes no sense to assume that we have a characteristic, an innate intelligence, which we can measure. It makes no sense because as Joanna Ryan said "in the process of measurement some aspect of the individual's current behavior has to be used, that is, some of the skills that develop during a lifetime."[8]

The fact that intelligence does not develop in a vacuum is not surprising, for this is true of any human trait. We are, as are all organisms, the product of the combined action of our genes and environment. Neither heredity nor environment alone can influence an organism's development in isolation from the other. To paraphrase the famous song about love and marriage, we really ". . . can't have the one without the other." It is true that some traits are more influenced than others by heredity or by environment, but any trait is influenced by both.

However, the idea that intelligence is in part hereditary is not the issue here. Nor is it how important heredity is to intelligence. What is important is to understand that although *individuals* differ in intelligence, this does not mean that *the groups wrongly called races* are differently endowed in this respect.[9] These two very different ideas are commonly found intertwined in the minds of most laypersons, journalists, newscasters, and even some scientists.[10]

Today most experts are well aware that intelligence tests do not measure innate intelligence. However, the belief held in the past that they did led to their misuse. For example, in the first forty years of the twentieth century, it was believed that people with low IQ had little innate intelligence and could not be educated to fill useful places in society. Children once stamped as "retarded" were left in the hands of educators who believed they could not be educated, and so did not try to educate them.

The same idea was applied to immigrants to the United States whose IQ scores were low and therefore judged to be mentally defective. Such judgments of individuals were extended to the ethnic group to which they happened to belong and the whole group was presumed to be inferior. In fact, a possible reason for their low score was that they did not know the intricacies of American language and culture. To prevent the influx of supposedly inferior people into the U.S., it was suggested to Congress that it pass laws preventing the immigration of such undesirable people.

The Congress obliged in 1924 by passing a law not only restricting the total number of immigrants to be admitted, but also assigning quotas on national origin. As a result, millions of people, especially those from southern and eastern Europe, were prevented from ever seeing the Statue of Liberty. It is also this law which led ultimately to the death of thousands of German victims of the Nazi regime. They were denied entry to the United States because the "German quota" was filled, though other quotas were not. Fortunately today this Immigration Act has been repealed. Today, the idea that IQ tests measure innate intelligence is still believed by some people and is used to buttress the idea that human groups such as "whites" and "blacks" differ genetically in intelligence. Early in the twentieth century, it was discovered that, in general, "black Americans" scored lower than "white Americans" on IQ tests. This discovery led to very heated, even incendiary, debates because two different and seemingly opposite hypotheses had been offered to account for the findings. One was that "blacks" differ genetically from "whites" in intelligence, while the other ascribed the differences to environmental influences, low socioeconomic standards, poor schools, discrimination, and other factors that we do not know about. One of the most famous debates about race and intelligence stemmed from an article written in 1969 by Arthur Jensen, professor of educational psychology at the University of California at Berkeley. This article, published in the *Harvard Educational Review*, was entitled " How Much Can We Boost IQ and Scholastic Achievement?"[11] It created a furor in the United States, especially among those in the Civil Rights movement, because Jensen attributed the IQ score differences between American "black" and "white" students predominantly to hereditary differences. One of his conclusions was that programs such as Head Start, established to raise scholastic achievement, were just wishful thinking and that tax money spent on them was wasted. This particular conclusion made him very unpopular with the disadvantaged. Jensen's position brought down on him a storm of personal abuse for political reasons.

In addition, geneticists and psychologists heavily criticized him and pointed out the many errors in his arguments. It was these errors that caused Jensen's conclusions to be rejected by most of his fellow scientists, not the unpopularity of his ideas. If he had been scientifically correct, scientists would have defended him in the name of academic freedom even if they did not like his conclusions. However, he was not scientifically right and they told him so. In spite of all the criticism, Jensen continued to defend his position by writing numerous articles and books. He was not alone in his views. In fact he was supported by Richard Herrnstein, a professor at Harvard University, who wrote an article in the magazine *Atlantic* in 1971.[12] This aroused another storm of controversy because Herrnstein attempted to link intelligence not only with race but also with social class. English psychologist, H. J. Eysenck not only shared Jensen's view, but believed as Herrstein did, that there were differences in intelligence between social classes, and that these differences also were in great part genetic. These men were highly vocal and did a great deal of harm because, being professors at prestigious universities, they were assumed to speak with real authority.

In the last few years, however, their influence has been weakened as a result of attacks by prominent psychologists and geneticists who banded together to fight this pseudoscientific racism. At the 1989 annual meeting of the American Association for the Advancement of Science in San Francisco, Philippe Rushton, professor of psychology at the University of Western Ontario, presented his views on "racial" differences in intelligence. He claimed that "Mongoloids," or natives of the Orient and their descendants, had evolved more recently than "Caucasoids," who in turn had evolved more recently than the "Negroids." Rushton's views were blasted by many of the scientists present at the meeting. They said he had completely misinterpreted the data of others and employed ideas that had been known to be wrong for one hundred years. Despite this criticism of his ideas, Rushton wrote a book that was published in 1994, in which he stated (without proof) that races differ not only in brain size—hence according to him, in intelligence—but also in reproductive physiology and anatomy.[13] Also in the fall of 1994, a book, *The Bell Curve: Intelligence and Class Structure in American Life*,[14] written by the late Richard Herrnstein and Charles Murray was published.

These authors have once more tried to link intelligence to race. Within a few days of publication, *The Bell Curve* was discussed and heavily criticized in many popular magazines, as well as on radio and television.[15] The book is a rehash of arguments that have been presented previously and shown to be erroneous. It employs faulty logic, accepting data which support the authors' views while rejecting data which do not.[16] The authors also drew faulty conclusions to suggest changes in the way that society deals with the poor. In summary, people who ascribe the differences in IQ scores of "black" and "white" Americans or other "ethnic groups" to hereditary differences in intelligence are on shaky grounds because we are not sure what IQ tests measure; we have not been able so far to define intelligence; and we cannot use IQ tests devised for members of one culture for members of another culture. But they are also on shaky grounds for another very important reason, which most critics of Jensen, Rushton, Herrnstein and Murray have missed until today.

In almost all of the studies carried out to test differences in intelligence (as measured by IQ tests) between "black" and "white students, it has been assumed that they belonged to two distinct biological "races" that can be easily recognized. This is not historically and biologically so. As we have mentioned previously, there have never been pure races. Matings between "whites" and "blacks" have occurred since before the beginning of recorded history. Hence, the "black" slaves who were imported to the New World, or those who came later from the West Indies had some "white" ancestry. But even if we assume that blacks and whites were two different biological populations before the importation of slaves into America, this would not be true today. Matings between the two populations have occurred on a regular basis. The most prevalent type was between slave owners or their sons with black women. However, frequent matings also occurred between black slaves and white indentured servants. A lesser known instance of mating occurred between white women and black men in the New England Colonies where, for a while, there were fewer white men than women.

Black-white matings also occurred during the Civil War and its aftermath. It is difficult to determine what the immediate effects of the freeing of slaves had on the sexual relations between whites and blacks. It is possible that these relations increased because blacks were now free to move to the cities where they were in contact with whites. However, there is a consensus among sociologists that sex relations between members of the two groups has decreased in the twentieth century. In all, these cases of black-white matings, the offspring were considered by American whites as "blacks" according to the infamous one-drop (of blood) rule. Only a person with 100 percent white ancestry was and still is considered to belong to the white race. Such a concept of race is, of course, a cultural one, not a biological one.

The amount of white ancestry in the American black population is far from being insignificant. It has been possible for geneticists to approximate it. If we could assume that there existed a black population and a white population at the time slavery began in North America, each of which carries only its own genes, it would be possible to calculate the fraction of "white genes" in the American black population and also the fraction of "black genes" in the American white population. Geneticists have made this assumption implicitly but not explicitly, and have calculated that, on the average, one fourth of all genes that the black population carries are "white" and not "black." On the other hand, on average, one out of every hundred genes that the "white" population carries is a "black" gene. The presence of "black" genes in the "white" population is explained by the fact that some children of black parents have skin pigmentation, hair type, and facial features which are similar enough to those of whites to "pass" for whites. If these individuals married white individuals, they would transmit "black genes" to their "white" children.

The fact that the American black population has a large amount of white ancestry is ignored by race classifiers who classify people primarily on the basis of skin color and only secondarily on other visible traits. Margaret Mead[17] puts it this way:

> The social classification of Negroes . . . is based on some known, often minute, attribute or visible amount of Negro ancestry. It ignores all ancestries, whether European or Asian, so that Negroes are treated sociologically and politically as if they are a race . . . It ignores, too, the simplest logic of genetics, which should attribute equal weight to paternal and maternal lines, recognize that individuals receive their particular genetic characteristics from their particular parents and not from a population . . . .

What Margaret Mead deplores is what was done in almost all of the studies carried out to test differences in intelligence between black and white students: The so-called racial background of the individuals has been derived on the basis of the subject's self-classification or by someone else in authority, either the teacher or the investigator. In the rare instances where information is given about the criteria used for racial classification, one usually finds that skin color has been the sole or dominant criterion.

Once classified, the student is given an IQ test and the examination scores are then analyzed according to the assigned racial classification of the student. The design of these studies is based on the premise that skin color is a good indicator of how much black ancestry an individual has. This premise is a bad one for the following reason. Let us continue to assume that there once existed a black population and a white population, each of which carries only its own genes, including different genes for intelligence. The science of genetics teaches us that, within a hybrid population, such as the U.S. black population, genes for traits such as intelligence and skin color are shuffled like playing cards in the course of reproduction. In other words, there is no reason, a priori, to link any other characteristic, including intelligence, with skin color. The little, very dark-skinned girl who scored 200 on an IQ test[18] may have or may not have inherited white genes. This view is supported by some studies. Attempts to show that students having the largest amount of white ancestry scored higher than those with the least amount of white ancestry have failed.[19]

The correlation between test scores and physical traits indicating greater or lesser black ancestry is so weak that it is doubtful that there is any physiological mechanism linking those traits and doubtful that the genetics of intelligence and skin color are in any way related. Nevertheless, the assumption that black and white students belonged to two distinct biological groups was so anchored in the minds of those who compared the IQ performance of these two groups that they never asked if two white (or two black) populations could differ from each other as much in IQ scores as one black and one white population do. They never designed their experiments to answer this fundamental question. Yet, such an answer can be found in studies that were made in Israel[20] and in France.[21] Two principal different ethnic groups came to Israel: one from Western Europe, the other from North Africa, the Middle East, and India. The first group, from western Europe, was more educated than the other. Years ago, IQ tests were given to preschool children of both groups and it was found that the children of the first group surpassed those of the second group by eleven to sixteen points on those IQ tests. Such a difference is similar to the one that has been found between "white" and "black" students. All of these Israeli children were put in kibbutzes, where they received the same education away from their family environment. Three years later they were again given IQ tests. At that point in time, the IQ scores were the same for both groups. More recently, French educational psychologists also showed that IQ scores could be raised many points by changing the social situation of the child at an early age. Children of unskilled workers who were adopted by professional families performed as well as the biological children of the professional families, and had an increase of fourteen points over their siblings who had been reared with their original families.

The results of the above experiments indicate that the differences in IQ among the children are due to environmental factors—yet when the same differences involved black and white children, they are ascribed to their heredity. Logic would require experimenters to ascribe the same cause to both types of situations. But they do not. The reason is simple. They assume that the black and white children belong to different biological groups. This assumption is widespread not only in IQ studies but in medical studies,[22] where the consequences might be as catastrophic. In conclusion, if the objective of the studies carried out to test differences in intelligence among black and white students was simply to learn whether or not there are differences in IQ scores among them, there is nothing to criticize. But to attribute any difference found to different genetic endowments is unfounded. The differences in IQ test scores that one can demonstrate between black and white Americans show only that under present cultural conditions in the United States, those persons who are regarded as black do worse on certain types of examinations than those persons who are regarded as white.

Since IQ and human races are human inventions and not naturally existing things, i.e., nonentities, an idea first expressed by Ashley Montagu in 1975 (see the epigraph of this chapter), and more recently by Alan Wolfe[23]—where does that leave the race-IQ debate? The IQ-race question is a pseudoscientific problem that is kept alive seemingly for political and social reasons.

Notes for Chapter Fourteen

1. First paragraph of the introduction of *Race and IQ*, edited by Ashley Montagu, Oxford University Press, 1975.

2. Richard Herrnstein and Charles Murray, *The Bell Curve: Intelligence and Class Structure in American Life* (New York: The Free Press, 1994).

3. In his book *Cyril Burt, Psychologist* (Ithaca, N. Y.: Cornell University Press, 1979), L. S. Hearnshaw, describes the life of an eminent psychologist who committed the scientist's cardinal sin: deception and fraud.

4. Alfred Binet and Theodore Simon, "Méthodes nouvelles pour le diagnostic du niveau intellectuel des anormaux," *Annales psychologique* 11:191-244 (1905).

5. Winston Groom, *Forrest Gump*, (New York: Pocket Books, 1986).

6. The following example demonstrates how a person's background can influence his or her performance on an IQ test. A child from a poor Kentucky family was presented the following problem: "If you went to the store and bought six cents' worth of candy and gave the clerk 10 cents, what change would you receive?" The child replied, "I never had ten cents and if I had, I would not spend it for candy, and anyway, candy is what your mother makes." The intelligence tester tried again, reformulating the problem as follows: "If you led ten cows out to pasture for your father and six of them strayed away, how many would you have left to drive home?" The boy replied, "We do not have ten cows, but if we did and I lost six, I wouldn't dare go home." The intelligence tester made one last effort. "If there were ten children in a school and six of them were out with the measles, how many would there be in school?" The question was no sooner asked than it was answered. "None, because the rest would be afraid of catching it, too." It is obvious that no inference about the mathematical ability of this Kentucky child can be drawn from this type of test. Yet many tests have been indiscriminately applied and far-reaching conclusions later found to be wrong have been made about the intelligence of children.

7. R. L. Williams, "The Bitch 100: A Culture Specific Test." Paper presented at the eightieth annual convention of the American Psychological Association. Honolulu, September 1971.

8. Joanna Ryan, "The Illusion of Objectivity," in *Race and Intelligence: The Fallacies Behind the Race-IQ Controversy*. Ken Richardson and David Spears, eds. (Baltimore: Penguin Books, 1972).

9. The best discussion of this subject can be found in the article by Richard Lewontin, "Race and Intelligence" in the March 1970 issue of *Science and Public Affairs, The Bulletin of Atomic Scientists.* This article has been republished in many books, for example, in *Race and IQ* edited by Ashley Montagu (Oxford University Press. 1975). See also the article by Jerome Kagan, "Inadequate Evidence and Illogical Conclusion," Harvard Educational Review 39:274-277 (1969).

10. One of them was Arthur R. Jensen, who attempted to demonstrate (through faulty arguments) that intelligence has a genetic component. "How Much Can We Boost IQ and Scholastic Achievement?" *Harvard Educational Review* 33: 1-123 (1969).

11. Jensen, *Harvard Educational Review* 33: 1-123.

12. Richard Herrnstein, "IQ". *The Atlantic* 228: 43-64 (1971).

13. J. Philippe Rushton, *Race and Evolution and Behavior: A Life History Perspective* (New Brunswick. N. J.: Transaction Publishers, 1994).

14. Herrnstein and Murray, *The Bell Curve: Intelligence and Class Structure in American Life* (New York: The Free Press, 1994).

15. A two-hour televised symposium was held at Howard University on December 12, 1994. Members of the panel dismissed the book as flawed.

16. The authors of *The Bell Curve* wrote the two following paragraphs on p.106:

> If, one hundred years ago, the variations in exposure to education were greater than they are now (as is no doubt the case), and if education is one source of variation in IQ, then, other things being equal, the heritability of IQ was lower than it is now.
> This last point is especially important in modern societies, with their intense efforts to equalize opportunity. When heritability rises, children resemble their parents more, and siblings increasingly resemble each other; in general, family members become more similar to each other and more different from people from other families.

In these two paragraphs Hernstein and Murray managed to contradict one of their most important assumptions relative to the nature of IQ, to use faulty logic, and to give the impression that they know very little about heredity.

In the first paragraph, they state that education is one source of variation in IQ. This statement is true, and contradicts one of the main assumptions about IQ that Hernstein and Murray make in their introduction (p. 22-3), that IQ scores are stable over much of a person's life. If education indeed increases IQ scores, then it is logical to believe that the poor test scores of children from poor school districts are due to inferior education, not heredity. And it is logical to assume that improving these inferior schools will be beneficial in improving the IQ performance of these students. These conclusions run contrary to the main theme of the *Bell Curve*.

In the same paragraph, we read: "If one hundred years ago, the variations in exposure to education were greater than they are now,...and if education is a source of variation...then heritability of IQ was lower than it is now." The last statement does not follow from the first two. The only thing we can conclude from the first two is that the variation in IQ must have been greater one hundred years ago.

Faulty logic is also found in the second paragraph: "When heritability rises, children resemble their parents more . . . ." This is not true. What is true is that if children could resemble their parents more for some elusive reason, heritability would increase. But the reverse is not true. It seems that Herrnstein and Murray confuse heritability, which is a mathematical estimate of the amount of variability in a trait due to the genes, with heredity, which is the concept that individuals related by descent resemble each other more than individuals who are not related.

17. *Science and the Concept of Race*, Mead, Dobzhansky, Tobach and Light, eds. (New York: Columbia University Press, 1968).

18. Curt Stern, "The Biology of the Negro," *Scientific American* 191:81-85 (1954).

19. Audrey Shuey, *The Testing of Negro Intelligence* (New York: Social Science Press, 1966). Also Sandra Scarr and Richard A. Weinberg, "Intellectual Similarities Within Families of Both Adopted and Biological Children" *Intelligence* 1:170-191 (1977).

20. Jerry Hirsch, "Behavior-Genetic Analysis and Its Biological Consequences." *Seminars in Psychology* No. 2, pp.89-105. Republished as *The IQ Controversy*, J. N. Block and Gerald Dworkin, eds. (New York:. Pantheon Books, 1976).

21. R.C. Lewontin, Steven Rose, and Leon Kamin. *Not In Our Genes* (New York: Pantheon Books, 1984), 128.

22. Medical researchers do not question the reality of racial categories. Each year, dozens of reports in journals use them to show differences between "races" in the susceptibility of diseases, infant mortality rates, life expectancy, blood pressure, and a host of other pathologies. This use of racial categories tends to lead us to believe that there are genetic differences in health among races, when it is highly possible that these differences are caused by environmental factors that can be easily detected.

23. Alan Wolfe. "A Response to the Bell Curve," *The New Republic* (31 October, 1994): 17.

Chapter 15

RACE AND DISEASE: ANOTHER PSEUDO-PROBLEM

Race is a small but volatile word, it lacks a clear definition or scientific purpose, yet it persists not only in the lingo of the streets but in the language of the laboratory.

James Shreeve [1]

We have seen in previous chapters that the assumption that human races exist has led science astray for quite a while. Unfortunately this assumption still persists unchallenged today in the field of medicine. While the physical differences among us, such as shape of nose and hair, are not important, our differences in susceptibility to diseases and our likelihood of inheriting a genetic diseases may be of great significance.

Medical researchers, unlike anthropologists, have not yet questioned the use of racial categories as a means for organizing data. They use such techniques to argue that there are clear differences among "racial" and "ethnic" groups in susceptibility to diseases, infant mortality rates, life expectancy, and other markers of public health. The readers of these reports are led to believe that the differences in health among human groups are genetic in nature when they are most likely due to environmental and socioeconomic causes.

Many authors of articles and books that deal with racial/ethnic differences in the study of disease are aware that the use of the concept of race is, at best, imprecise. They constantly warn their readers to be aware that economic and cultural factors, and not genetic factors, may be responsible for these differences.[2] Nevertheless, without telling us why, they continue to use their racial groupings, hoping that, in some undefined way it will lead to a better understanding of diseases and how they work. However, it is difficult to understand how racial classification, which has been shown to have no explanatory value in understanding human variation, could be a useful tool in finding causes, diagnoses and treatments for human disease.

We have demonstrated that biological differences between human beings are of degree and are not absolute; that is why it is impossible to classify mankind into races. Scientific information indicates that this also applies to the study of humans and disease. Though certain diseases affect some populations more than others, the same kinds of disease affect all of humanity.

The continued use of racial groupings in medical research is therefore harmful because the term "racial" implies something that has been taken for granted but never demonstrated, namely that there are highly significant biological differences between certain groups of people. Even the United Nations has recommended dropping the word "racial" from its lexicon and using only the word "ethnic". "Ethnic" is strictly a social term. Ethnic groups may differ culturally; they may have different traditions, including religious beliefs. They may also have different personal habits such as diet, smoking, use of alcohol, sexual and reproductive patterns, including marriage patterns.

These differences may be maintained for long periods of time, even when different ethnic groups live together in the same country. Such variations in the ways they live their lives contribute to the different manifestations of diseases observed in human groups.

Differences in disease among ethnic groups may also reflect socioeconomic factors and discrimination. The dropping of racial categories and the use of well-defined ethnic terminology might lead to a better understanding of diseases and their prevention. In many cases medical researchers have confused racial groups which they assumed to exist, with ethnic groups which they have either never described or have described poorly. They have often included data from "racial" and "ethnic" groups in the same table as if they were interchangeable.[3] One also can find that the same population is treated as racial in one table and as ethnic in another. One can find examples where people having "Spanish" surnames are treated as an ethnic group in contrast to people with "non-Spanish" surnames.[4] Are we to assume that a woman whose maiden name is McGinnis becomes biologically different when she marries a man named Fernandez? It is obvious from these examples (and there are many others) that the reporting of data on group differences in disease suffers from a lack of precision, standardization, and common sense.

There are many studies that compare the health of American blacks and whites. These studies have indicated that there are differences between these two groups' susceptibility to disease. For example, black men are supposedly 40 percent more likely to suffer from lung cancer than are white men, and black women tend to develop tumors that are more malignant than those found in white women. Are these studies pointing to genetic differences between blacks and whites? It is unlikely. The same arguments that were presented in chapter 14 on race and IQ can be used here. Blacks and whites are not distinct biological groups. Matings between the two groups have occurred for centuries. To assume that they carry different genes for susceptibility to disease is to assume that these genes are associated with skin color. There is no evidence to demonstrate such a link.

However, even given the very poor quality of our classification of mankind into groups, it still cannot be denied that there are differences in frequencies of diseases among human groups. The cause of these different frequencies cannot be ascribed to fundamental biological differences between these groups, but rather to sociocultural and environmental conditions under which these human groups live.

The above statement applies to most diseases. But there are diseases which are far more common in some ethnic groups in than others and which express themselves regardless of sociocultural and environmental conditions. I am referring to genetic diseases. Today, we know a lot about genetic diseases. We know how they are inherited and how they affect their victims. Though we do not know for sure why they happen to occur more frequently in one particular human population than in another, we have a good idea that these diseases originated by mutation of a specific gene; that they became established in a given population due to local customs and conditions; and that they remain within a particular ethnic group because of the tendency of the members to marry within their groups. We also know that the same disease could have occurred in another ethnic group under favorable circumstances. In other words, contrary to what many people believe, these genetic diseases are not specific to a particular human group because that group is inherently biologically different from any other group, but because the disease happened to have originated in one population and not in another.

For example, there are genetic diseases that affect Jews (more precisely, one specific population of Jews) than any other human group. Tay-Sachs disease comes to mind. This disease is about ten times more common among Ashkenazic Jews than in other populations. The term Ashkenazic is applied to Jews of central and eastern Europe and to their descendants. Tay-Sachs disease manifests itself soon after birth, usually within four to six months, killing the child by the age of two. In this degenerative disease, complex fatty substances accumulate in the nerve cells, including those of the brain. This leads to a progressive degeneration of cerebral functions, resulting in blindness, convulsions, paralysis, idiocy and eventual death. The disease is due to a defective gene. The normal gene is responsible for production of an enzyme that degrades fats. This enzyme is not present in Tay-Sachs victims and is responsible for the symptoms of the disease. It has been suggested, but not proven, that individuals who have a normal gene and one defective were favored in the special environment of a ghetto, where food was not always abundant. They were favored, supposedly, because they did not break down fats as fast as normal people and therefore could draw on body fats in case of food shortages.

Whatever the reason why Tay-Sachs prevails among Ashkenazic Jews, it is not because Ashkenazic Jews and their descendants are inherently biologically different from other Jews or other human populations. This view is supported by the following fact: A disease clinically identical to Tay-Sachs disease occurs in Catholic French Canadians in Quebec, but the defective gene seems to be slightly different than that found in Jewish populations.[5] This suggests that the gene responsible for the production of the enzyme that degrades fat can mutate in slightly different ways, giving similar symptoms of the same disease.

No human population is free from any specific genetic disease. There is a myth that cystic fibrosis is strictly a European disease, found primarily among Irish people and their descendants. It was believed to be nonexistent in Japan. However, since the 1950s several cases have been reported in that country.[6] The same was true for another genetic disease, Glucose-6-phosphate dehydrogenase deficiency (G6PDD) which is discussed later in this chapter. Seven Japanese males were found to have the deficiency.[7]

There is a medically very important genetic disease that most people associate exclusively with one particular racial group. The disease is sickle cell anemia and the "racial" group is the African blacks and their descendants. That the disease occurs only among this group is a myth that is perpetuated by journalists (and other writers) who called sickle cell anemia the "black disease." To understand why this is a myth, we have to explain the nature and the origin of this disease.

Sickle cell anemia is the result of an abnormal form of the protein hemoglobin. This protein, located in the red blood cells, has an important function: the transportation of oxygen to every cell of the body. In most of us, our red blood cells remain circular in shape even when deprived of oxygen. But in those of us who are affected by sickle cell anemia, the red blood cells change shape when deprived of oxygen. It is this change of shape that is called "sickling." There are three kinds of people in respect to sickling. There are those who are normal; none of their cells sickle. There are those who are said to have the sickle cell trait; half of their cells are normal and half of them abnormal. They live normal lives except when they are at high altitudes or any other situation where the amount of oxygen available is reduced. Finally, there are those who are said to have sickle cell anemia; all of their cells sickle whenever a large amount of oxygen is needed. When this occurs, the patients have excruciating pain due to the blood clots caused by the accumulation of "bunched sickle cells." Their life expectancy is short, for the disease claims many victims by the time they are twenty years old, the rest by the age of forty.

Sickle cell anemia is a genetic disease. Individuals with sickle cell anemia have two defective genes for the disease, and individuals with the sickle cell trait have a good gene and a defective one. The defective gene produces a protein that differs from the normal protein by the substitution of only one amino acid for another, but this slight difference is great enough to cause havoc with the blood cells of the victim. The abnormal hemoglobin is less able to be dissolved than the normal hemoglobin and tends to crystallize, resulting in the deformation of the red cells. The abnormal fragile cells age faster than the normal ones and die faster than they can be replaced (they survive only seventeen days instead of the usual one hundred days).

The sickle cell gene reaches a high frequency (sometimes as high as 20 percent) in Africa, among human populations extending from West Africa to Madagascar. For a deadly gene to attain a frequency as high as 20 percent is very remarkable, since the victims die before they reproduce. There must be a logical explanation.

The explanation came from the observation that the frequency of the sickle cell gene in different populations was correlated with the frequency of malaria in the same groups. Malaria is one of a family of diseases, the varieties of which affect many mammals and birds. The parasite responsible for malaria is carried from one victim to another by mosquitoes and therefore has always been more prevalent in areas of humid climates, where the mosquitoes are continuously active throughout the year. Malaria has been a major cause of death throughout human history. In Africa today, malaria is endemic, that is, it does not sweep through a population as an epidemic, but rather is a constant affliction contributing to early childhood mortality rates as high as 50 percent. It kills about 10 percent of its victims and contributes to the death of others by decreasing the ability of the immune system to fight other infections. The immune reaction to the malaria parasite is incomplete, and though the disease is usually contracted during childhood in malaria-infested countries, adults may be reinfected.

The fact that the frequency of the gene for sickle cell anemia was high in areas where malaria was prevalent suggested that malaria had something to do with it. It was known for a long time that people with the sickle cell trait were less susceptible to malaria than people who did not have a gene for sickle cell anemia. However, it took a long time to discover why. Finally, it became clear that while the parasite responsible for malaria grew equally well in red cells in normal persons, in persons with sickle cell trait, or in individuals with sickle cell anemia as long as oxygen was abundant, this was not true when there was a shortage of oxygen. In the latter case, while the organism grew very well in the blood cells of some people, it did less well in the cells of the persons with sickle cell trait, and did not grow at all in the cells of persons with sickle cell anemia.

Hence, in areas where malaria is prevalent, there is a definite advantage in being an individual with sickle cell trait. Such an individual does not get malaria and is not affected by sickle cell anemia. Since he or she has one normal gene and one abnormal one, the frequency of the gene for sickle cell anemia will remain high.

The fact that "African blacks" and their descendants have the genetic disease sickle cell anemia has nothing to with the color of their skin, shape of nose or hair, but is due to the fact that they were living in areas where malaria was prevalent. If white people had lived in these areas, malaria would have played the same role for them and we would have associated sickle cell anemia with white people instead of black people. As a matter of fact, there are white people who suffer from sickle cell anemia in Arabia, India, and parts of southern Europe, where malaria was, and in some cases still is, prevalent.

This view is also supported by the fact that malaria has been responsible for the high occurrence not only of sickle cell anemia but of other genetic diseases, such as thalassemia and glucose-6-phosphate dehydrogenase deficiency. Thalassemia, a blood disorder, resembles sickle cell anemia in its genetics, distribution and medical effects. This heritable disease which is also known as Cooley's anemia, or Von Jackh's disease, is found primarily among Mediterranean people and their descendants. But it is also found in Asia Minor, India and some parts of Indonesia where malaria is also prevalent and where individuals who have one normal gene and one defective gene are at an advantage.

In children affected by the severe form of thalassemia, the red blood cells are small and abnormal and the hemoglobin differs in just one amino acid, just as in sickle cell anemia. Because the cells lack the normal hemoglobin, they are unable to carry enough oxygen and are subject to rapid destruction in the spleen, which enlarges because it has more than the normal number of cells to destroy. Within the first year of life, infants with thalassemia are noticeably anemic and fail to thrive. Infections are frequent. They have subnormal growth and die before the age of puberty. Parents of such children have red blood cells that differ slightly from the normal shape.

Glucose-6-phosphate dehydrogenase deficiency is another example showing that genetic diseases are not confined to one particular human group. It is widespread around the world, where it also represents a genetic adaptation to malaria similar to those of sickle cell anemia or thalassemia. Though there are many forms of G6PDD due to many variants of the defective gene, they affect the individual in the same way. The anemia has been traced to a deficiency of an enzyme which plays an important role in ridding the red blood cells of foreign particles. G6PDD affects millions of people from the Mediterranean region to Asia and Africa. Their red cells are affected in the presence of certain drugs, such as primaquine,[8] insecticides such as moth balls,[9] or foods such as the fava bean.[10]

Though most of the differences in susceptibility to disease among human groups are unlikely to be genetic in nature, they do exist, and medical researchers should continue looking for the true causes of these differences. Knowledge of these differences is of extreme importance to clinicians in providing the best care to their patients. In some cases the ancestral origin of the patient may suggest the probable diagnosis of a disease. For example, fever and abdominal pain may have different implications to a northern European (who is possibly suffering from appendicitis), a Mediterranean (possibly favism or familial Mediterranean fever), or an American of African ancestry (possibly sickle cell crisis). In the last two cases, the patient may be spared needless surgery. Though it is useful to know the family/ ancestral background of the patient, the physician should treat him or her as an individual.

Notes for Chapter Fifteen

1. James Shreeve, "Terms of Estrangement," *Discover*. Special issue: The Science of Race. Nov. 1994. p. 57.

2. Anthony P. Polednak, *Racial and Ethnic Differences in Diseases* (New York: Oxford University Press, 1989).

3. Polednak, p. 174

4. Polednak, p. 95

5. Polednak, p. 81

6. Henry Rothschild, *Biocultural Aspects of Disease* (New York: Academic Press, 1981), p. 314.

7. Rothschild, p.315.

8. During World War II, it was found that the drug primaquine, which was used to prevent malaria crises among the soldiers, precipitated blood crises mostly among the black GIs. About 10 percent of the black American males were found to be affected. The disease is more prevalent in males than in females because the recessive gene responsible for the deficiency of the enzyme is borne on the X chromosome. Because males have only one X chromosome instead of two, the chances that they have the disease are far greater than their female counterparts. It appears that the deficiency caused by the defective gene confers some protection against malaria. And this is the irony of the situation: to prevent malaria, primaquine was administered to individuals who probably would not get the disease in the first place, and the drug ended up killing or seriously affecting them by causing a G6PDD crisis.

9. The severe reaction to moth balls observed in some children who play with them is tied to Glucose 6-phosphate dyhydrogenase deficiency. Naphthalene, the compound in moth balls, destroys the red blood cells which have the defective gene.

10. Pythagoras, the famous Greek philosopher, had warned his students not to eat fava beans or even walk through a field of fava beans. Was he affected by favism, the disease characterized by the acute destruction of red blood cells, after eating fava beans or inhaling the plant's pollen? It is highly possible, for favism is widespread along the Mediterranean Sea and affects many individuals.

Chapter 16

HOW THE U.S. GOVERNMENT CLASSIFIES ITS CITIZENS: A REAL PROBLEM

"I yam what I yam"

Popeye, the sailor man.

If scientists have for the most part abandoned the concept of race, the U.S. Government, like the public at large, has not. Today, the question, what race do you belong to? is asked by all types of institutions in the United States, because the federal government wants to know if they are complying with the famous Article Nine, which deals with affirmative action. Affirmative action is a phrase that refers to attempts to bring members of underrepresented groups, usually groups that have suffered discrimination, into a higher degree of participation in some beneficial program.[1] To achieve this, the government has to be able to recognize the members of the groups who have suffered in the past. Hence, it asks everyone to choose his or her race/ethnic group from a set of categories that have been determined by one of its agencies, the Office of Management and Budget.

The importance of this classification cannot be overemphasized, since these racial/ethnic categories are present on all federal forms and application forms for jobs, scholarships, loans, mortgages and so on. The federal classification is utilized not only by most public and private affirmative action programs, but also in surveys by the Equal Employment Opportunity Commission, the Department of Education, the Department of Labor or federal contract compliance programs, and by all colleges and universities in their mandatory biennial reports on personnel and salaries, enrollment and degrees.

Following are the most recent federal definitions of racial/ethnic categories as used in the United States:

1. White (not of Hispanic origin): all persons having origins in any of the original peoples of Europe, North Africa, the Middle East, or the Indian subcontinent.

2. Black (not of Hispanic origin): all persons having origins in any of the black racial groups.

3. Hispanic: all persons of Mexican, Puerto Rican, Cuban, Central or South American, or other Spanish culture or origin, regardless of race.

4. Asian or Pacific Islanders: all persons having origins in any of the original peoples of the Far East, Southeast Asia, or the Pacific Islands. This area includes, for example, China, Japan, Korea, Philippine Islands, and Samoa.

5. American Indian or Alaska Native: all persons having origins in any of the original people of North America.

It is clear that this classification of American people is not based on scientific criteria, anthropological or other. It intertwines supposed "racial groups" and "ethnic groups." For example, "whites" and "blacks," which for years have been considered racial groups are divided into two "ethnic" subgroups, Hispanic and non-Hispanic. Why such a subclassification? Except for the language, a "white" man or woman who comes from, or whose family came from Central or South America cannot be distinguished biologically from a man or woman whose origin is Italian or Southern French. The reason for this subclassification is a social one, being concerned mainly with the fact that there has been and, as the author is personally well aware, still is discrimination against people whose geographic ethnic origins are in Mexico, Puerto-Rico, or Central or South America.

Any classification, whether of human beings, books, or rocks is artificial. However, contrary to rock classification which does not fundamentally change, the federal classification of the people of the United States has continuously changed as a result of lobbying by various special interests.

While the white category has persisted unchanged for a long time, the number and types of minorities has not. For example, in 1956 there was a large influx of Puerto Ricans into the mainland cities of the U.S. There they looked for jobs in the construction business. To ensure that contractors did not discriminate against them, the federal government asked contractors to report how many Puerto Ricans they employed. In other words, Puerto Rican became a minority category. But in time this category changed its name to Hispanics to include Mexicans when they became an important part of the labor force.

After Hawaii became a state in 1959, its congressional representatives influenced the Congress to recognize another minority group, namely the Orientals. Hence, the minorities became : Negro, Spanish-American, Oriental, and American Indian. However, the government soon ran into trouble also with its classification of Orientals. Polynesians, Hawaiians and the Guamanians in Southern California did not consider themselves as belonging to the same group as the Japanese or the Chinese. Hence, the Oriental category was divided into two categories, the Asian and the Pacific Islander. This brought another problem. For years immigrants from India, Pakistan, and Sri Lanka had been classified as "whites." But in 1977 they requested to be considered as a minority so they could be included in affirmative action programs. From that year on, they were considered Asians, that is they are in the same category as the Chinese and Japanese. More recently, the Arab American Institute asked that persons from the Middle East, who had been counted as "whites," be given minority status.[2]

It has been pointed out by many that the Federal classification of its citizens is full of contradictions. For example, if a descendant of the original people who lived in North America before its "discovery" by Columbus lives north of the Rio Grande, he or she is considered an American Indian, but if that individual lives south of the River, he or she is considered Hispanic. People who immigrate directly from Cuba to the States are considered to be "Hispanic," but if they immigrate to the States after having passed a few years in Brazil, they are considered white.

It has also been pointed out that many people cannot be neatly put into any one of the categories that are offered to them. For example, where does a friend of mine, who has a French father and mother who herself has an American Indian father and a Venezuelan mother, place himself? Is he white? American Indian? Or Hispanic? According to federal regulations, he can choose his category. However, if he does not choose and does not mark any category box, he will subsequently be required to choose a category or someone else will mark a box for him judging only from his physical appearance.[3] This raises two important questions. First, how accurate can the data be if obtained in such a fashion? Second, how seriously do the U.S. government officials take this classification?

The answer to the first question is that the data are not accurate, as was found by Robert A. Hahn, an epidemiologist at the Centers for Disease Control and Prevention. He discovered that the reason for the staggering increases in the infant mortality rate for minority populations in 1985 was due to the fact that the infant had frequently been given a different race on its death certificate from that on its birth certificate.[4]

Another example of misinformation is the following. Ron Moskow-itz, education correspondent for the *San Francisco Examiner*,[5] re-ported in 1974 that many white families had found an original way to prevent the bussing of their children far away from their homes. They declared their children to be American Indians or blacks. They can do that, since guidelines issued by the U.S. Department of Health, Edu-cation and Welfare say school districts must accept whatever racial or ethnic background parents give them. The only thing the school dis-trict officials can do is to prevent parents from changing the race of the child back at will, once declared. The result of this was that, in that particular year, the San Francisco school district was faced with an unusually large numbers of American Indians, blacks, Koreans and Japanese.

The answer to the second question, how seriously does the U.S. government regard these data, cannot be given with any certainty. As an example, years ago, one of my colleagues who wanted to find out called the local U.S. Census Bureau for information. Being a good citizen, he said, he wanted to fill out the census form he has just re-ceived through the mail as accurately as possible. His problem was that his father was Chinese, his mother English, his paternal grandfa-ther American Indian, his paternal grandmother Mexican, his mater-nal grandfather Armenian, and his maternal grandmother Egyptian. The local representative of the U. S. Census Bureau told my friend to call the state representative. The state representative told him to call Washington. Finally, my friend got the associate director of the U.S. Census Bureau who, after listening to him told him: "Who cares! Put anything you want."

My friend's ethnic origin was, of course, a product of his fertile imagination. But consider the true case of a professional actor, Teja Arboleda. His father's father is Filipino-Chinese, his father's mother is African-American and Native-American, his mother's father is Dan-ish, and his mother's mother is German. He calls himself the "ethnic man." Pointing at his face, he asks: "Can you honestly say, without disgrace, there is such a thing as pure culture or pure race? I rest my case."[6]

The fact that the U.S. guidelines are not based on precise criteria and the loose way that questions about race and ethnic groups can be an-swered has not only prevented any meaningful statistical survey or analysis, but it has also brought anxiety and pain to individuals who do not want to be or cannot be pigeonholed into any of the given ra-cial/ethnic categories. I am referring to individuals whose parents "belong" to two different categories. In this case, the U.S. Govern-ment forces the individual to choose, ignoring, or wanting to ignore, a fundamental fact of nature, namely that an individual is the product of two parents.

The failure of the government to deal with people of "mixed origins" is not accidental. American society in general has denied that there are such people. A child from a black parent and a white parent is considered black. Children of Asiatic and white parents are classified as Asiatic. The one-drop rule, or some form of it, remains the law of the land. Only a few years ago, the U.S. Supreme Court let stand a ruling that a woman "whose fraction of black ancestry was only 3/32" was legally black.

Recently, a multiracial classification has appeared on all school, employment and state forms in Ohio, Illinois, North Carolina, and Georgia. A fifth state, Michigan, is considering establishing such a multiracial category. This may help destroy the one-drop rule and helps the children of mixed marriages who do not want to betray either parent. Superficially, it may help us to get a better picture of human diversity. But, in fact, it perpetuates the myth that there were or are pure races and that a child can be a mixture of two of them. A child, as I have emphasized in this book, is a unique individual who should not have to be categorized and who should be able to say, like Popeye, "I yam what I yam" and be happy. If these categories were to be removed, we all could be happy like Popeye, and this may be about to occur as early as the year 2000. The House Subcommittee on Census, Statistics, and Postal Personnel is considering not only modifications of existing racial/ethnic categories but also the larger question of whether it is proper for the government to classify people according to arbitrary distinctions of skin color and ancestry.

Notes for Chapter Sixteen

1. Kent Greenawalt, *Discrimination and Reverse Discrimination*. (New York: Alfred Knopf, 1983).

2. Lawrence Wright, "One Drop of Blood," *The New Yorker*, July 25, 1994

3. The U.S directive is the following: "An employer may acquire the race/ethnic information necessary [to report] either by visual surveys of the work force, or from post-employment records.

   It is recommended that visual surveys be conducted for the employer by persons such as supervisors who are responsible for the work of the employees or to whom the employees report for instructions or otherwise."

4. Wright, "One Drop of Blood."

5. *San Francisco Examiner*, 9 June 1974.

6. As reported in the *Detroit Free Press Magazine*, 9 Oct. 1994.

INTRODUCTION TO PART 3

In parts 1 and 2 of this book I have attempted to develop the idea that there is no scientific evidence to support the assumption that human races exist. I have emphasized that this assumption has led scientists astray for a long time. It is only in the last few years that scientists have really begun to reverse their thinking. By abandoning the idea that human races exist, they are now in a far better position to look at human diversity

In part 3, I am giving a modern view of what species and races are; I emphasize again why human races do not exist; I explain why each of us is unique; I develop the idea that the differences in skin color between us are not absolute, but relative and subject to change.

Chapter 17

OF SPECIES AND RACES: A MODERN VIEW

> Genera, orders, classes exist only in our imagination . . . There are only individuals. Nature does not arrange her words in bunches, nor living beings in genera.
>
> *Buffon*, Histoire Naturelle

> "All men are created equal" but Negroes were said not to be men.
>
> *Gunnar Myrdal* [1]

In the eighteenth century, opponents to slavery in Britain's American colonies accused slave owners of not considering blacks to be human beings, but as members of another *species*. This accusation was not entirely deserved. Though most planters treated slaves as animals, they did not necessarily believe that they were not human beings; among them, Thomas Jefferson comes to mind.[2] But, of course, it also must be remembered that pro-slavery writers wrote explicitly about Africans not being of the same species as whites.[3] The pro-slave polemicists went so far as to place blacks in groups that included orangutans, apes, baboons and monkeys.[4] Such a ridiculous classification method was applied to avoid their addressing a fundamental moral issue. By denying that black slaves were human beings they also denied them the rights of humans. By using the word "species" and other scientific terminology, pro-slavery writers tried to give a scientific luster to their explanations of why African slaves should be kept in bondage. "Species" is indeed a scientific term. The word comes from a modern classification scheme that was devised to bring order to the great variety of organic life which shares the planet Earth with us. It comes from Latin and, put plainly, means "kind," or type. We see around us all "kinds" of plants and animals.

Walking through a pine forest, we are aware not only of the stately pines but also of other trees, such as the cedars and scrub oaks. In some parts of the forest, we can see squirrels gorging themselves on acorns and pine seeds, while deer crop leaves from the succulent vegetation. We can see hawks hovering overhead, ready to swoop down with outstretched talons to scoop up an unsuspecting rabbit. Pines, cedars, oaks, squirrels, hawks, rabbits are kinds of organisms that are easy to distinguish. But, further observation leads to the discovery that there are different kinds of pines, or that there are different kinds of squirrels. White pine, red pine, pitch pine and jack pine all share features that indicate a close relationship and that permit us to group them together and be named pines. They are evergreen trees with spreading branches, with needlelike leaves, with male and female flowers, with cones that contain winged seeds. Yet they all differ among themselves. For example, white pines have a dark gray bark and bluish green needles, while pitch pines have a reddish brown bark and light green needles.

Animals have similar patterns of relationship. Various kinds of squirrels have much in common besides the way they gorge themselves on acorns or pine seeds. They also have many differences, not only in size and color but also in the type of habitat they occupy. For example, in open woodlots one might find fox squirrels with their black faces. In the dense parts of a pine forest, one can find red squirrels, which are smaller than the fox squirrels and do not have black faces.

The system now in use to classify plants and animals was started more than two hundred years ago by the Swedish naturalist, Carl von Linné. In his system, each kind of living thing was given two names. For example, the proper name for pet dogs is *Canis familiaris*. It was Linné who gave domesticated dogs that name and named wolves *Canis lupus*. *Canis*, the genus name, includes other species, besides *familiaris* and *lupus*. Linné considered dogs and wolves similar enough to be grouped together. Just how similar various species must be for taxonomists (those who classify living things) to include them in the same genus is a matter of judgment. For example, again according to Linné, all wood frogs belong to the species *Rana sylvatica*. *Rana* is the name of the genus to which all the frogs belong. The wood frog bears the name *sylvatica* (silva means forest in Latin). The bullfrog looks like the wood frog and can be easily recognized as a frog, so its genus name is also *Rana*; but it belongs to another species because it is far bigger than the wood frog.

A similar classification system is used in botany. The name *Acer saccharum* designates sugar maple trees. They have leaves, flowers and fruits which identify them as members of a small group belonging to the genus *Acer*. The name *saccharum* also indicates that this plant belongs to a particular species, from which sugar is obtained. The pine trees mentioned above also have scientific names—*Pinus strobus* for white pine, *Pinus rigida* for pitch pine.

Defining a species in Linné's day was very different from what scientists do today. In the eighteenth century it was assumed that species did not and could not change. Each species, still in existence or extinct, was thought to have retained the same characteristics that it had when it was created. Accordingly, each species represented a *type* of organism, the variations of which were ignored or minimized. Hence, to describe a given species, all one needed was a few representative specimens. Linné classified mostly plants and animals of the Old World. However, as new lands were discovered during the voyages of exploration, the existence of new kinds of organisms was revealed. Only a few specimens were brought back to Europe. This small sample determined whether or not the organism was merely a subdivision of an existing species or was different enough to be called a new species. The scheme of classification was first designed simply as a method of pigeonholing species to help understand the range of living things and the relations between them. Organisms were classified according to what was called their morphology, that is, only according to their structure and form. But the French biologist Buffon (1707-1788), a contemporary of Linné's, first grasped the inadequacy of such a basis for classification. He pointed out that to compare the resemblance of individuals was only one of the tools for classification. Another one was the continuity of the species by breeding. He wrote:

> For the donkey resembles the horse more closely than the poodle resembles the greyhound; yet, the poodle and the greyhound are of the same species, since together they produce individuals which in turn produce others. The donkey and the horse, on the other hand, are certainly different species, since the individuals they produce are sterile.

There are numerous examples that show that the two criteria, morphology and breeding, i.e., mating and producing fertile offspring, are sometimes in conflict. One of them has to do with dogs, wolves, coyotes and jackals. These four kinds of animals have been classified as distinct species: *Canis familiaris*, the dog; *Canis lupus*, the wolf; *Canis latrans*, the coyote; and *Canis aureus*, the jackal. However, these animals can and do occasionally breed among themselves.[5] For example, during the late 1950s, hybrids between coyotes and wolves appeared in New England. It is believed that the hybrids developed because of a shortage of mates in a region where both species were practically exterminated.[6] In fact, all these four kinds of animals are morphologically distinguishable from one another. No wolf can be taken for a coyote or a jackal. But since the coyote and the wolf can have fertile offspring, they should be classified as being of the same species, which should include the dog as well as the jackal. But, in order to recognize their morphological differences, they should be regarded as subspecies of *Canis*, and we should give each a different subspecies name. In other words, *lupus*, *latrans*, *aureus*, and *familiaris* should be subspecies names and not species names. Though the breeding criterion presents certain difficulties in some instances, it has now become, and for good reason, the most important one for distinguishing species. Species differs from all other taxonomic categories in that it exists as units in nature independently of the wish or judgment of the classifier. If of all the members of a population can breed and have fertile offspring, they all belong to the same species. If not, they belong to different species. Unless we are sterile, all humans on the planet, regardless of who our mates might be, can have fertile children. That is why all of us, regardless of our physical appearance, belong to the same species, *Homo sapiens*. This being the case, we can see how wrong it was to consider "Negroes" as belonging to a lesser biological species.[7] If we are all members of the same species, can we be classified into different smaller groups which are commonly called races? Before we can answer this question, we must ask what biologists mean by the word "race."

Although we may not be aware of it, we all have an excellent example available to us of what biological races are: the different breeds of dogs we see every day. These range from the dainty little Chihuahua to the giant St. Bernard, from the curly Poodle to the Great Dane. How did these different kinds of dogs come into existence? They have been produced by us from some earlier form by selection of desired characteristics and restricted breeding. They would not have come into existence without our intervention, and each race would eventually return to the wild type from which it came if our controls were removed and it was free to breed at will. The wild ancestor from which our modern dogs were developed were very likely the Asiatic or Indian wolves, which were smaller than the large northern wolves. Even now, there are marked similarities between dogs and wolves, some of which echo throughout literature in such stories as Jack London's novels, *The Call of the Wild* and *White Fang*.

Though we can distinguish the dog and the wolf by their howling and barking, we know that each can learn to imitate the other. Both the wolf and dog wag their tails in pleasure and tuck them between their legs in fright; both can curl lips into a snarl when angered. Both use scent marked runways as a means of path recognition. Tooth and bone structure are similar in the dog and the wolf. Dog and wolf interbreed readily.[8] But both can also cross with the coyote and the jackal, with whom they also have many similarities.[9] While these crosses are infrequent in nature, they do occur and the offspring are fertile. The biological reason for this is that the dog, the wolf, the jackal and the coyote not only share the same number of pairs of chromosomes (39),

but in addition, if we take one of the pairs of dog chromosomes, we will find a matching pair in size and shape in the coyote, the wolf, and the jackal. This is necessary for successful matings. These animals have the same gestation period, 60–63 days, and they attain sexual maturity at the same age. However, there is a difference in the length of time from one receptive period to the next in the females: seven or eight months for the dog; one year for the wolf, the coyote, and the jackal. How did we shape the wolflike ancestor of the dog into descendants having so many different forms and qualities: a hunting dog such as the English setter, for instance; a sheep or cow dog such as the border collie; a guard dog such as the Great Dane; a police dog such as a German shepherd; or an ornamental dog that does absolutely nothing, such as the Yorkshire terrier? "Elementary, my dear Watson." We did this by selective breeding for hundreds of generations. Though no one recorded, of course, the happy moment when a "dog" first licked the hand of a human being instead of biting it, we can estimate that this domestication took place 10,000 to 12,000 years ago, about the time of the end of the last great ice age.

At first, when *Homo sapiens* were strictly hunters, we kept only the dogs most useful in the hunt. This was a primitive type of selective breeding. When we domesticated sheep and cattle, we needed dogs to ward off predators, fast-running dogs that were intelligent and obedient. How did we get them? Obviously, at first no dog had all of these qualities. But within the whole population of dogs at the time, there were some which, though not obedient, ran fast. Others were obedient but slow. Others were very intelligent, but not necessarily fast and obedient. Our ancestors thought about a way to produce dogs with all of these three qualities. How about mating the ones who run fast with ones who are obedient; hopefully, some of their puppies might be obedient and swift, though they might not run as fast as their father and might not be as obedient as their mother. Nevertheless, on the whole, these puppies would be useful. These dogs would be even more useful if they were very intelligent. Mating these dogs with Ginger, a bitch who is remarkably smart, should give puppies that would be fast, obedient and smart. Of course it took many more generations of selective breeding than is suggested here to get Lassie the sheep dog. But this, in outline, is how we have obtained not only Lassie but all races of dogs, including the robust and fierce dogs that guard our homes and police our cities, the fast-running dogs that are able to outrun the quarry we hunt, and all the breeds of show dogs and toy dogs. All of these modern races of dogs will remain intact only as long as dog breeders keep them separated from one another. This isolation is absolutely necessary.

What would happen if a door was suddenly opened in the wall of isolation maintained by dog breeders? Think about your favorite dog, for

which you have opened the front door to let it go outside to answer a call of nature. It may encounter a dog of the opposite sex and a different race. Sixty-three days later, you may be presented with a litter of puppies sharing some of the characteristics of your dog and some of the characteristics of the other contributor. Because of this mixture of characteristics, the puppies cannot be classified as members of either race. This is an example of what would happen if selection pressure were relaxed. Hence, selection and isolation are keys to the formation and maintenance of races in any organism.

In nature there is no such strict selection or isolation as can be found in any of our plant or animal breeding programs. However, there is selection brought about by the environment, and there are isolation mechanisms. These are not as effective as those provided by the plant or animal breeder in creating new races. Though there are no fences or cages in nature separating populations of organisms, there are natural barriers to prevent widespread mating among them. Distance is an important factor. In plants, pollen does not travel very far. Animals do not cross a whole continent to find a mate.

But isolation can also be of a different nature. To understand this, let us take two kinds of snails. They may not be able to mate because they are on opposite sides of a sea, a lake, a mountain range, or a freeway. Though they live in the same area, one prefers a moist soil and the other a dry one; or they are not sexually attracted to each other; or they do not breed at the same time; or they cannot physically mate, or if they can, they are incapable of producing viable offspring. In all of these cases, the two populations of snails remain distinct and in time, if isolation is maintained, will become very different. This evolution is slow. At the beginning, though geographically isolated, these two snail populations may be able to mate and produce fertile offspring if given the opportunity. Biologists would classify them as races. However, as soon as these two populations of snails lose the ability to breed among themselves, they have become different species. From this, it should be clear that the formation of races is a step on the way to the formation of species. In nature, different races of the same species simply evolve through isolation of various populations of the same species. The more isolated the populations are, the more they will differ in time from one another and qualify to be called races. More often than not, populations are not entirely isolated, and some breeding occurs.

Consequently, there are not always sharp lines of demarcation between races of organisms in nature as there are between domestic breeds of animals. Let us go back to our best friend, the dog. Every spaniel looks different from any golden retriever. Any Labrador looks different from a foxhound. In other words, we have no problems in distinguishing races of dogs. But we cannot distinguish races among

ourselves because there are no human races in the sense that there are races of dogs. The reason for this is clear. In the case of the dog, artificial selection and complete reproductive isolation were key ingredients to the production of distinct races. But human beings, with rare exceptions, have not been bred and selected by a super authority, or any other means, to produce special races of people. The closest we came to human beings under such authority, in recent history at least, was under Hitler, who started a breeding program in which young women and soldiers were selected to create a "super race" destined to populate the Third Reich for 1000 years. Fortunately, this program failed. It was not intensive and did not last much beyond four years.

Within certain boundaries, human beings have chosen their mates. In some cases, some of us have traveled thousands of miles to find a spouse, sometimes thanks to the generosity of our governments when engaged in military action. Wars have always offered wide opportunities for breeding between the invaders and the invaded. But even in peacetime, attractions between the sexes has often been enhanced by differences in culture, customs, and manners of the two individuals involved. Sometimes it even overcame the lack of common language. For reasons of this kind, we were never isolated in the past, we are not isolated today, and most likely we will not be isolated in the future. The process of race formation, which was highly successful in the case of dogs and other domesticated animals, is continuously nipped in the bud in the case of humans because no group of us is ever reproductively isolated from any other.

Notes for Chapter Seventeen.

1. Gunnar Myrdal, *An American Dilemma: The Negro Problem and Modern Democracy* (New York: Harper and Brothers, 1944).

2 According to Winthrop Jordan (*White Over Black*, p. 431-2), Thomas Jefferson never for a moment considered the possibility that blacks were rightfully enslaved; they had God-given rights because they were human beings. For a detailed account of Jefferson's ideas about the subject, see Daniel J. Boorstin, *The Lost World of Thomas Jefferson* (New York: Henry Holt and Co., 1948), pages 81-108.

3. See William Stanton, *The Leopard's Spots: Scientific Attitudes Toward Race in America* (Chicago: The University of Chicago Press, 1960).

4. See Winthrop Jordan, *White Over Black: American Attitudes Toward the Negro, 1550–1812* (Chapel Hill: The University of North Carolina Press, 1968), pp. 301-305.

5. The defenders of slavery knew that dogs and wolves interbreed. However, since these animals were considered to be two different species, slavery proponents used this fact to argue that though whites and Negroes could have fertile children, they were not of the same species. See Jordan, *White Over Black.*

6. See Stephen J. Gould, "What Is A Species?" *Discover* 12 (December 1992): 40-44.

7. According to Ginsberg and Eichner, only the conception of the Negro as an inferior species can explain the roster of statesmen seeking to remove the Negro from the United States' shore.

8. Since 1787 we have known that the wolf, the jackal, and the dog are the same species. See J. Hunter, *Philosophical Transactions of the Royal Society* 77 (1977): 253-266.

9. See Roy Robinson, *Genetics for Dog Breeders* (Oxford: Permagon Press, 1982).

Chapter 18

EACH OF US IS UNIQUE

. . . each sperm or each egg receives a copy of half the information that had initially been transmitted to this individual by his parents at conception, and based on which he was gradually constructed.

Albert Jacquard[1]

Most Catholic, Jewish, and Protestant groups teach their congregations to accept people on their individual worth regardless of their name, color, religion, or occupation. This moral advice is not only excellent but agrees with what biology has taught us for years. We are each biologically unique; the basic reason can be summarized in a few words. The incredible diversity that exists among us is the result of one fundamental aspect of sexual reproduction.

Each of us has a unique genetic makeup, a unique set of genes. The chances of our having such a genetic makeup can be compared to our chances of having the winning ticket in a national lottery. This is because, contrary to what is commonly believed, each parent does not transmit all of his or her genes to a child, but only half of them. The two halves are then juxtaposed, without being mixed, to constitute a complete set of genes, which is really a collection of different pieces of biological information. However, and this is the clue to our diversity, each half of the biological information that comes from the father or the mother varies with each child. Hence, each child is the result of a specific different combination of genes present in his or her parents. The number of these combinations is greater than the number of electrons in the universe.

Each child is, therefore, not a reproduction of anyone, but a definite, unique creation, due to chance. The reader might be astonished to find the word "chance" in the language of science. After all, science is supposed to deal only with *true* and *certain* knowledge. Scientists try to understand how observed phenomena are interrelated and then use this understanding to predict future phenomena. Thus, an astronomer having understood the movement of planets and the attraction between masses, can accurately predict solar and lunar eclipses. This knowledge also can apply to our understanding of the movement of satellites around planets.

However, there are outcomes of natural phenomena that scientists have been unable to predict. For example, they cannot predict which of the six faces of a single die will be on top once it is thrown. This is because such a throw involves too many unknown factors, such as the characteristics of the die, the strength of the throw, and the type of surface on which it falls. We say, therefore, that the outcome, which face will be on top, is left to chance. In a similar fashion, scientists are unable to predict which genes will be present in a child because there is no way to know which sperm among the millions present in a sexual act will fertilize the egg, nor which genes will be present in that sperm. Nor do they know which of the mother's genes will be present in the egg. Because they are unknown, we say that the inherited characters of each child are due to chance. In this case, our predictive ability is, at best, limited. Only for a few traits, under very specific circumstances, can we predict with accuracy the characteristics that the child will have. With some other traits the best we can do is to estimate the probability that a child will have a specific characteristic. With most traits we are unable to predict a thing. For example, there is no way of knowing how smart or long a newborn is likely to be.

Each of us is unique for two reasons. Although we have the same number of genes, the form of the genes differs from one person to another. Furthermore, we grow and live in different environments. We are influenced by the environment from the day of our conception: first, as we develop in the womb of our mothers, then as we develop in an exterior milieu after birth.

Today, we believe that both heredity and the environment play parts in the expression of all of our traits, though in some cases one may appear to be more important than the other. In practice, characteristics are found which are little affected by the usual variation in external conditions, such as blood types and fingerprints. But most other characters are very greatly influenced by the environment. Farmers are well aware that even their best dairy cattle will produce milk that is poor both in quantity and quality if she is underfed, and the best wheat seed will yield only a poor harvest if it is grown on sand. We are affected by our environment in similar ways. For example, chronic malnutrition during pregnancy and during the early years of life is known to bring a drastic reduction in the number of brain cells present in human beings. This brain damage has been estimated to affect millions of children throughout the world, children who are unable to compete effectively in school or for jobs.[2]

Though it is now generally understood that each character or property of an organism is the product of numerous innate and environmental influences, it was thought at an earlier time, and it is still thought in some circles, that any characteristic could be explained either by inborn or by environmental influences, but not by both. Quite a few years ago, a former professor of mine asked this question on an exam: "Is this particular trait due to heredity or to the environment?" One of the students answered: "What the hell kind of a question is this?" The professor gave him nine points out of ten, because he realized that the student understood perfectly that no trait was due simply to heredity or to environment, but was the result of the interaction of both. But he took one point off for insolence. Another question the professor could have asked is "What are the relative roles of heredity and environment in determining a particular trait?" However, such a question is as bad as the first, because we cannot assign separate causal roles to internal or external forces in the formation of individuals. We have seen in a previous chapter that people ask this very question regarding intelligence and how the asking of this question has kept alive the pseudoproblem of "race" and intelligence.

We can compare the roles of heredity and environment to the exposure and development of a photographic film. We load a camera with color film and make some exposures, yet we have no visible image on the film. There is a potential image, however, which can be brought out by developing agents. If we have made a good exposure and the proper chemicals are used for the proper time and temperature in the development of the film, we have a good color transparency. Even with the best exposure, however, poor development will not realize the potential. Likewise, if we make a poor exposure, no amount of good development can make a good transparency. In the case of development of organisms, heredity gives the potential, and environment functions like development. We shall illustrate and discuss the interaction of these two important factors later in this book.

In the next three chapters, I will present the modern view of heredity and discuss some of its consequences, including one that will undoubtedly upset a large number of people who are proud of some of their remote ancestors. The importance they attach to ancestry rests on two fallacies, namely that heredity was a process of passing blood from parent to child, and the other, that blood carried factors that influenced character. Touches of genius, of great courage, of brilliance-or taints of criminality, of shiftlessness and depravity—were thought to be carried in the blood. Genetics has demonstrated clearly that we have inherited very few genes, if any, from our remote ancestors.

In conclusion, our individuality is due in great part to a process where two principal players, our parents, each provide us with a collection of copies of only half of their genes. These collections, which vary with every child that the couple has, are the result of chance. Since the number of possible collections is practically infinite, we can assume that no one was, is, or will be identical to us (except for those of us who have an identical twin). Furthermore, our individual diversity is not only the result of inheriting different biological information from our parents, but is also due to the fact that in growing up no two of us had the same life experiences. The fact that each one of us is unique makes nonsense of our exaggerated tendency to judge people according to the social or "racial" groups to which they may be allocated.

Notes for Chapter Eighteen

1. Albert Jacquard, *In Praise of Difference: Genetics and Human Affairs* (New York: Columbia University Press, 1984).

2. See Elie A. Shneour, *The Malnourished Mind* (Garden City, N. Y.: Anchor Books, 1975).

Chapter 19

OF GENES AND CHROMOSOMES: NO ONE IS LIKE YOU

> He does not realize that, instead of conceiving him, his parents might have conceived anyone of a hundred thousand other children, all unlike each other and unlike himself.
>
> *Peter Medawar*
> Nobel Prize Winner

In chapter four, we mentioned that the thrilling movie *Tainted Blood* was based on three false assumptions, one being that there was something "special" about a boy and a girl who were twins. But the twins, in this instance, were of different sexes and were therefore fraternal twins. Fraternal twins, unlike identical twins, are no more similar than regular brothers and sisters. They originate from separate fertilization events: two eggs, ovulated in the same menstrual cycle, each fertilized by a different sperm. Different eggs and different sperm combine to produce complete and different individuals, which happen to be born the same day. Identical twins, on the other hand, originate from a single fertilization event, a single egg fertilized by a single sperm. This fertilized egg, a special cell called a *zygote*, accidentally divides at an early stage in its development to form two separate cells each of which grows into a separate embryo. In this case we have two individuals with the same heredity and obviously of the same sex.

Before explaining further why different eggs or sperm have different hereditary makeup, we must say a few words about genetics, a science of heredity that appeared within the study of biology about 1900. Genetics revolutionized scientific thinking about how traits are transmitted from parents to offspring. According to this science, elementary units of inheritance, which are separate and independent of one another are passed with mathematical precision from one generation to the next through the sex cells. It is these units, called *genes*, that are inherited—not "traits," and certainly not "race," which is, in fact, a construction of our minds.

It is in the course of human development that genes, in interaction with the environment, produce our physical characteristics. This view contrasts with the old idea that the characteristics of parents were blended in the offspring, as one might blend two liquids by mixing them together, as discussed in chapter four. Gene theory, on the contrary, asserts that there is no blending or dilution of individual characteristics. To press the analogy, genes produce either clear water or water of a particular hue. Moreover, genes will continue to create these distinct products generation after generation.

Different genes have different functions. For example, one gene is responsible for the production of the growth hormone. Another is responsible for the production of the beta chain of hemoglobin, the protein that carries oxygen to all cells of our body. These two genes not only have different functions, but they are different in their internal structure.

Though genes are exceptionally stable in composition and function, they can and do change through a process called mutation. Genes reproduce themselves and it is during the process of self-replication that mutations may occur. If a gene happens to mutate to a new form, subsequent generations of that gene reproduce themselves in the mutant, or altered, form. Such alternative forms of a gene are known as *alleles*. The extreme diversity that we observe among human beings is due to the fact that we have different alleles, not to the fact that we have different genes. And this is true for any species.

Alleles lead to the different forms of various products that genes manufacture. For example, we all have wax in our ears. Some of us have wet wax, others have dry wax. Some of us have both kinds of wax. The different types of ear wax are due to different alleles of the gene responsible for producing that substance.

Though geneticists themselves have defined the terms gene and allele, they have, in many cases, used the two words interchangeably. For example, Curt Stern wrote in his book, *Human Genetics*:[1]

> It is to be expected that American Caucasians possess African *genes* acquired by direct race mixture and by the process of blacks "passing" into the Caucasian population. Numerically, this African fraction among Caucasian *alleles* is small. It has been estimated, with 95 percent probability, that it is less than 1 percent.

Stern should have used the word allele throughout the paragraph. There is no such thing as an African gene, but there might be alleles more common among Africans than in other populations. As we are all human beings, we all have the same genes. If we had different genes, we would not be human beings; we would be something else— chimpanzees, cats, or whales. It has been shown that we have 98 percent of our genes in common with the chimpanzees.[2] The rest must represent the different genes that make us two different species.

What makes us different from one another is the fact that we have different alleles. The fact that some of us have darker skin is not due to our having different genes, but because we have different alleles. The same is true for height, eye color, or any other trait that we have. This distinction between genes and alleles should be kept in mind because we are going to use these words throughout the rest of this book.

When I was a teenager, I read a French novel whose title has long been forgotten, which contained a discussion I will always remember. The author wanted to stress the point that the chances of his having come into existence on earth were small. To illustrate his case, he declared that a few minutes before his conception, his mother asked his father if he had wound the clock.[3] He remarked that if his mother has not done so, he would not have been around to write his novel. His parents would have had a child, but it would not have been him. This French novelist knew his biology. The egg that produced him would have been the same in any case since a woman has only one egg a month, but men produce millions of sperm. Therefore the chances that a particular sperm would fertilize the egg are very infinitely small. Since each person is the result of the fertilization of one particular egg by a particular sperm, the birth of our novelist, it could be argued, was almost a pure accident, and so is the birth of each of us.

But what is it that makes each sperm unique, i.e., different from all others. To understand, we must turn again to our genes. Genes are located along microscopic threadlike bodies called chromosomes, which are themselves located within the nucleus, a dense body generally at the center of each cell. These chromosomes are found not only in the sex cells but also in all the body cells of an organism. It is the chromosomes, with the genes that they carry, which we inherit from our parents rather than blood.

In every species, the number and shape of the chromosomes are constant and generally different from those of other species. We have twenty-three pairs of chromosomes, with one of each pair coming from each of our parents. In twenty-two of these pairs, called *autosomal* chromosomes, each chromosome is similar in shape to the other. But in the twenty-third pair of the male the two chromosomes are not similar to each other. These chromosomes making the twenty-third pair are the sex chromosomes and are called X and Y. The Y chromosome is far smaller than the X chromosome and contains fewer genes. A woman has two X chromosomes, while a man has an X and a Y chromosome.

In our body, new cells arise through division of old ones. The original cell divides into two new cells. Within the original cell chromosomes duplicate and each of the two new cells receive forty-six chromosomes (twenty-three pairs) identical to those contained in the original cell. This process by which a cell reproduces into two identical cells is called *mitosis*. Mitosis is involved in the production of new cells in the growth of new tissues or in the repair of existing ones.

But there is another cell division that is responsible, in large measure, for the fact that each of us is genetically unique. During the formation of our sex cells, a process called *meiosis* reduces the number of chromosomes by half and in such a way that each sex cell is genetically different from all others. In meiosis, sex cells are the product of two consecutive, special cell divisions in which the forty-six chromosomes, after they have duplicated, are distributed so that each of the four new cells receives only one chromosome from each pair, twenty-three chromosomes in all. Each egg produced by a woman has the twenty-two autosomal chromosomes and one X chromosome. Sperm also have the twenty-two autosomal chromosomes but there are two types of sperm, one with an X and one with a Y chromosome. At fertilization, the union of egg and sperm forms a new cell in which the characteristic number of chromosomes (forty-six) is restored once again. The embryo will be male if the egg is fertilized by a Y sperm, female if fertilized by an X sperm.

We have stated earlier that the genes, the units of inheritance, are located along the chromosomes. Each gene occupies a certain place on a chromosome. This place is called a *locus*, and since chromosomes occur in pairs, each gene has its counterpart at the same position on the corresponding, or *homologous*, chromosome. These two genes can be identical or be of two different forms. As I have mentioned already, the different forms of the same gene are called *alleles*. The fact that genes can have different forms makes the chromosomes of each pair different from each other. Chromosomes have their own individuality.[4]

Now that we have seen why chromosomes have their own individuality, it is important to understand how we, as human beings, acquire our individuality. Each of us received half of our chromosomes from our father and half from our mother. But, though your brother and sister have the same parents as you, chances are that they did not receive the chromosomes identical to the ones you received from your parents. Your father received twenty-three chromosomes, called *paternal chromosomes*, from his father and twenty-three chromosomes, called *maternal chromosomes*, from his mother. However, because each of his sperm—through the process of meiosis—contains only one chromosome of each pair (which one is determined at random), the sperm that fertilized the egg of your mother to produce you did not receive the same combination of chromosomes as the sperm which fertilized the egg of your mother that produced your brother or your sister. The same can be said for the sex cells of your mother. Any sex cell from your parent could have received either fifteen paternal chromosomes and eight maternal chromosomes or thirteen maternal chromosomes and ten paternal chromosomes, to name just a couple of possibilities. Since we have twenty-three pairs of chromosomes, there are 2^{23} or 8,368,608 possible combinations for a person's gametes, assuming that only two alleles, two forms of the same gene, govern one trait and that no other factors influence the process. The chances that you had received the same combination of chromosomes from your parents that your sibling received from them is the product 8,368,608 by 8,368,608, which for practical purposes is infinity. This means that two children of any couple of parents have no chances whatsoever of being alike. We can say that any human being is the result of a unique combination of chromosomes, hence is genetically unique.[5]

The fact that each of us is unique is of great significance and cannot be overemphasized. We are each the product of a unique heredity and a unique environment. There has never been anyone like either of us, and no one will ever be like either of us. To a biologist, one of the most remarkable features of the human population is its enormous variety. Here among six billion members of one species, we find no duplicates. Literally, each person is biologically unique and declares this fact not only in his or her obvious physical features but in the individual properties of his or her proteins, in the operations of his or her sense organs, and in numerous details of chemical constitution and behavior.

This individuality should be understood, respected and cherished. Every person must be judged according to his or her individual qualities, regardless of how we might classify them.

Notes for Chapter Nineteen

1. Curt Stern, *Principles of Human Genetics* (San Francisco: W. H. Freeman, 1973), p. 832.

2. This extreme similarity between us and chimpanzees has been supported by the work of molecular biologists who used special techniques, one of them being DNA hybridization.

3. The French novelist might have borrowed the idea from Laurence Sterne, the Irish author of *The Life and Opinions of Tristram Shandy*. On page 2 of the book one finds the following: "Pray, my dear," quoth my mother, "Have you not forgot to wind up the clock?"—"Good G—!" cried my father, making an exclamation, but taking care to moderate his voice at the same time, "Did ever woman, since the creation of the world, interrupt a man with a silly question?" Pray, what was your father saying? Nothing.

4. The individuality of each chromosome has permitted geneticists to pinpoint in an individual which chromosome of each pair came from his or her mother and which chromosome came from his or her father.

5. There is a phenomenon called crossing over—the splitting and rearrangement of chromosomes—which occurs during meiosis that decreases even further the chances that an individual has the same combination of chromosomes as another individual.

Chapter 20

MYTHS ABOUT ANCESTRY

Mr. Reginald Twombley Dunn-Twerpp—who is not very bright, weighs 110 pounds and is the first to climb on a chair at sight of a mouse—likes to boast that he is descended from William the Conqueror and that the steel-blue blood of ancient warriors flows in his veins. To prove it he will show you his family tree and a beautiful hand-painted crest, prepared by a genealogist in Boston for fifty dollars.

Amram Scheinfeld[1]

Many people, like Reginald, still believe that blood is involved in inheritance. According to this notion, the blood of our ancestors, which carries their characteristics, somehow mingles and is poured into us. Our inheritance is a half-and-half blend of blood from our fathers and mothers, and hence a quarter each from our grandparents, one eighth each from our great grandparents and so on.[2] This notion, a logical consequence of the blood theory of inheritance, is widely held. If believers of this theory have a famous man or woman somewhere in their lineage, they, like Reginald, boast of the fact, believing that they share some of his or her blood and consequently some of his or her remarkable qualities.

But the notion that all of our ancestors contribute their due share to our makeup is wrong. Though we realize that doting grandparents may not like to discard this belief, we will now attempt to destroy it. After all, grandparents are sure that they have contributed one-fourth of their biological makeup to their grandchildren. There are four grandparents, they reasoned, and each one must have contributed the same amount. It is not that simple, as we will show in the following pages. However, we are sure that, once they know the truth, the grandparents will continue to love their grandchildren as much as they did before.[3]

Did you ever think that you have a tremendous number of ancestors? You have two parents, four grandparents, eight great-grandparents, sixteen great-great-grandparents, and so on. This is correct since mathematically, the number of ancestors per generation increases by a power of two and can be represented by 2^n, n being the number of generations. Six generations back you have sixty-four ancestors, twenty generations back 1,048,576 ancestors. Forty generations back the number of ancestors becomes 10,955,116,000,000,000. Such a number is beyond our imagination. Now, since in all probability, the number of human beings on the entire planet did not exceed 425,000,000 until the year 1650, we soon reach a time in the past when the hypothetical number of ancestors we are supposed to have had outnumbers the actual number of people on earth. Hence, we are forced to realize that we all have common ancestors and therefore we are all related. This thought must shake you up, as it does me. I am happy to know that I am related to someone I admire, but I am distressed that I am related to someone I hate.

The reader will undoubtedly think: "O.K. So if we go far enough into the past we will find that we all are related. But I am more related to my great-grandparents than I am to someone who lived in Africa or Asia in the tenth century." Possibly, but not necessarily to someone who is your ancestor six generations back.

The reason why this is so is the same as the one which we gave for the fact that we are unique: the process of meiosis that occurs during the formation of sex cells. Though you can be sure you inherited twenty-three chromosomes from each of your parents, you cannot know how many chromosomes you indirectly received from your grandparents. As you remember from our previous discussion of meiosis, your father had received from his own father twenty-three chromosomes that we have called paternal chromosomes and from his mother twenty-three chromosomes that we have called maternal chromosomes. However, because his sperm contains only one chromosome of each pair (which one is determined at random), any one of them can contain any combination of paternal and maternal chromosomes; any one of them could have received either fifteen paternal chromosomes and eight maternal chromosomes or thirteen maternal chromosomes and ten paternal chromosomes, to name just a couple of possibilities. It could happen that the sperm of your father which fertilized the egg of your mother that produced you had only one chromosome or no chromosome at all from your paternal grandfather. It is highly improbable, but it is possible. In the same way it could happen that the egg that produced you had one or no chromosome from your maternal grandmother. Nevertheless, we can assume that, on average, we received eleven or twelve chromosomes from each of our grandparents, that an average of five to six came from our great grandparents, an average of two to three from our great-great-grandparents. With each generation further back, the average number of chromosomes we may have received from any ancestor is diminished by half. Consider now an important fact: Six generations back we have more ancestors than chromosomes (sixty-four versus forty-six). Hence, it is clear that the more remote our ancestor is, the greater the odds become that we did not receive even a single one of his or her chromosomes. That two individuals of the same family who are several generations apart are practically uncorrelated in their genetic constitution has also been demonstrated in a sophisticated mathematical way by a famous population geneticist, C. C. Li, who used genes instead of chromosomes.[4]

That there is practically no genetic link between someone and any of his or her ancestors six generations back is important, for it destroys some of the myths that surround family relationships. For example, the fact that my brother's stepson is descended from Benjamin Franklin makes an interesting item for conversation, but genetically it is meaningless. The relationship of this young man and Benjamin Franklin exists only on paper. This must be quite disquieting to people such as Mr. Reginald Twombley Dunn-Twerpp (but was not to my step-nephew), who are unduly concerned with pride in their ancestry.

That there is practically no genetic link between someone and any of his or her ancestors six generations back is also important in another way because it destroys some other social myths that surround ethnicity. One frequently hears a person says: "I'm one half-Irish" or "I'm one fourth-Indian" or "I'm one-eighth Jewish." Interestingly enough, no one says that he or she is one-fourth Protestant or three-fourths Catholic. These are recognized as religious labels, not as "racial" labels, while Jewish, which is equally religious-cultural, is assumed to be racial. All such references to someone being precisely this or that fraction of a given ethnic group have no meaning genetically for two reasons. The first is that, if ethnic or racial chromosomes existed (which I must emphasize *is not true*), the hereditary relationship of a person of "mixed ethnic" descent to any of his or her ancestral groups would depend entirely on the number of chromosomes of that group he or she carries. From the previous discussion, it should be clear that it is impossible to determine what such a number might be. For example, if a man who has a black father and a white mother marries a white woman, all of his children would be considered socially "one-fourth black." Each child having one and the same black grandfather is assumed to have received one fourth of his chromosomes. But actually any given "one-fourth black" child could be carrying any proportion, from zero to twenty-three of the supposed "black" chromosomes from his or her grandfather. Hence, it is possible for a person who is socially considered one-sixteenth black to be carrying more supposed black chromosomes than one who is considered one-eighth or even one-fourth black.

The second reason why the expressions above have no meaning genetically is that there are no Irish, French, or German genes, etc. The terms French, Irish, and German refer, after all, to nationalities. In the same vein, there are no "black" genes, but genes that affect skin color, shape of nose, hair form, and so on. It is these genes that we inherit, not ethnic group genes. What people—including some geneticists who should know better—call "black" genes are alleles of genes that affect skin color, shape of nose, blood types, and so on. As I mentioned before, some alleles may be more frequent in one group of human beings than in others. Alleles that are responsible for a darkening of skin color are obviously more frequent in equatorial Africa than in Sweden, Norway or Finland. But no one can tell from a person's skin color how many other so-called and not so obvious "black" alleles he or she is carrying, for example for blood types. And so a dark-skinned child may have fewer "black" alleles that a very light-skinned child. This is because the genes that determine skin color are only a small part of the total genes in a person and that these genes are, for the most part, inherited independently and work independently of other genes.

In spite of this, we still attach a social importance to our ancestry. We are proud of some famous ancestor who happened to be living in Ireland or equatorial Africa two hundred years ago. Of course, we ignore our ancestors who were hanged or jailed for stealing horses or cows, or, as a friend of mine was fond of saying, occupied a chair of applied electricity at a state institution. We do this because unconsciously we still believe in the blood theory of inheritance and ascribe to individual qualities of a whole group of ancestors. For example, we commonly explain someone's fighting spirit by saying such a person has gotten it from his Irish ancestors. However, such a trait may be "inherited" in strains of dogs and roosters by inbreeding and selection. These processes assure that each individual animal indeed has the fighting spirit. But we humans have always bred in a haphazard fashion and never selected for a particular trait. Hence, each of us, whether Irish or not, may or may not show a fighting spirit.

Notes for Chapter Twenty

1. Amram Scheinfeld, *Your Heredity and Environment* (New York: J. B. Lippincot, 1965), 635.

2. The prominent nineteenth century biologist Francis Galton was instrumental in transforming this belief into a "law," the law of ancestral heredity.

3. In many of today's families, including the author's, there are grandparents who love and spoil their nonbiological grandchildren (acquired through adoption or remarriage) and are proud of them as if they were biological grandchildren. And the kids do not mind it a bit.

4. C. C. Li, "A Tale of Two Thermos Bottles: Properties of a Genetic Model for Human Intelligence." In *Intelligence: Genetic and Environmental Influences*. Robert Cancro, ed. (New York and London: Gruen and Statton, 1971).

Chapter 21

OF DNA AND PROTEINS, OR NO ONE IS LIKE YOU

Except for identical twins, we are biologically unique.

Richard Lewontin[1]

We are alike, yet we are all different. If a Martian landed on Earth, he or she would have no problem in distinguishing humans as belonging to the same species. He or she would be struck by our erect posture which permits us to walk fast and steadily on the surface of the ground and to use our arms for things other than ambulation; he or she would possibly be impressed by our hair, which cascades from the top of our heads, but is almost absent from the rest of our bodies. These characteristics are only a few of those which make us similar.

If our Martian landed first in Sweden and then in Japan, he or she would be conscious that, though the people of these two countries are the same type of being, they are somewhat different in the color of their skin, hair and eyes; in the shape of their eyes and the structure their hair. If our Martian were allowed to stay a while in one of these two countries, he or she would be able, in time, to observe that no two individuals are really alike.

Where does this individual diversity comes from? In chapter nineteen I made the point that each of the chromosomes we inherit from our parents have their own individuality because they do not have the same genes, or more precisely, the same forms of genes. But this explanation, though correct, is incomplete. Our present knowledge of the structure and action of the genes has given us an excellent understanding of the nature of our individual diversity. Not only do we have different genes, but we have different proteins.

I have said previously that genes are responsible for the formation of enzymes, which are special proteins responsible for biochemical reactions in the cells. What is the exact relationship between genes and enzymes? To answer this question, we have to look at the general nature of genes and proteins. At this point, readers may wonder why we should dwell so deeply on the nature and function of genes when the subject discussed is race. Very simply, it is because knowledge of the nature and action of genes leads to a basic understanding of how the diversity of all organisms, including ourselves, can occur.

The field of genetics evolved from a pure abstraction to a real entity in less than seventy years. Evolution of such a concept is a tribute to the intensive and thoughtful work done by quite a few geneticists and biochemists. The concept of genes, the units of inheritance, was first used to explain data that breeders had obtained from experiments with hybrids. Later, genes were assumed to be located on chromosomes inside the nuclei of cells. This assumption has permitted us to understand, in great part, how genes were transferred from one generation to the next. Since then we have gone quite a few steps further. Not only have we succeeded in understanding how genes act, we have been able to locate many of them precisely and, recently, to transfer a few of them from one species to another.

Genes are situated on chromosomes. But where exactly? Chromosomes are composed of two different types of chemicals: nucleic acids and complex proteins. Ingenious experiments have demonstrated that it is the nucleic acid portion that contains the genetic information. A gene is part of a giant molecule consisting of a substance called *deoxyribonucleic acid* or DNA for short.[2] From a hereditary point of view, chromosomes can be considered chiefly as long threads of DNA.

DNA has a very important function. It serves as the blueprint that dictates which proteins are manufactured within cells. It is put together in such a way that it carries within itself a code that determines the order in which amino acids, the building blocks of proteins, are put together. There are twenty different amino acids. A protein may be made up of hundreds or thousands of these amino acids. Each protein owes its uniqueness to the specific amino acids it contains and the sequence in which they are put together.

DNA is a double-stranded molecule. Each strand is made of a long sequence of relatively simple compounds, the nucleotides, which are attached to one another. A nucleotide is made of the sugar *deoxyribose*, a phosphate, and one of the following bases: adenine, thymine, cytosine, or guanine. Since nucleotides are distinguished from one another only by the base, the four different types of nucleotides are identified by the letters A,T,C, and G, corresponding to the first letter of the name of each base. All these nucleotides occupy well-defined and ordered positions in a sequence which is specific not only for each species but for each individual organism. In other words, the fact that each of us has his or her own specific DNA sequence makes us biologically unique.[3]

The beauty of the DNA molecule is that the two strands of DNA are perfectly complementary, for whenever there is an A in one strand, there is a T in the other strand at the corresponding site. Opposite T is A, opposite G is C, and opposite C is G. The fact that A can only unite with T, T with A, G with C, and C with G permits an exact replication of DNA. Reproduction is achieved by the unwinding and separation of each strand, followed by replication of its missing complement.

It is these four bases that make up the genetic alphabet which directs the synthesis of proteins necessary for the cell's metabolism and development. We know that the English language, with its twenty-six letters which form thousands of words, can be expressed by a code composed of only two symbols: the Morse code with its dot and dash.[4] Hence, the genes can surely carry a very large amount of genetic information in a code composed of four symbols. Three letters in a code, a series of three bases, spells a single amino acid. A definite series of such three-letter words, arranged in a specific order on a section of the DNA chain, constitutes a sentence, a message. This message controls the formation of a complete protein or a complete chain of a multiple chain protein. The genetic message is indirectly translated because the formation of proteins does not take place in the nucleus of the cell where DNA is located, but in the material that surrounds it, the cytoplasm, and specifically to minute particles called *ribosomes*. The ribosome is a tiny yoyo-like object composed of two hemispheres of slightly different size. It is the ribosomes that receive the genetic information and use it to control the joining together of amino acids to form proteins. The genetic information is carried from the DNA to the ribosomes by a type of nucleic acid other than DNA. This other molecule is called messenger RNA. Like DNA, RNA is composed of a string of nucleotides. It is unlike DNA in three basic ways. First of all, the sugar in the RNA nucleotides is ribose instead of deoxyribose. Second, RNA is single-stranded, unlike DNA, which has two strands. And finally, there is no thymine in RNA; it is replaced by another base called *uracil*. Readers are referred to a genetics textbook if they are interested in the details of how RNA picks up the genetic message and transfers it to the ribosomes, where it is translated.

The self-replication process of DNA that occurs as one cell divides into two new cells is so exact that literally millions of self-replications can take place with perfect copies produced each time. Thus, through countless generations, the DNA making up a gene remains intact, guaranteeing the error-free transfer of genetic information from cell to cell and from parents to offspring. On very rare occasions, mistakes occur. One base may be substituted for another, with the result that a different amino acid will be inserted into the protein which is forming. Such a change in bases is called a mutation. Some of these mutations may have no effect or little effect on protein function. Many, however, alter the chemical structure in such a way as to produce vital changes in the organism.

This has been shown to be the case with sickle cell anemia, which affects many individuals among black Americans and Africans and a few other individuals as well (See chapter fifteen). Hemoglobin S differs from hemoglobin A, (the blood protein that carries oxygen to every cell of the body) only in the substitution of one amino acid for another, valine for glutamic acid on one position of the chain. This change of only one amino acid in a very large protein molecule is enough to change the structure of hemoglobin A into an abnormal hemoglobin that does not carry oxygen effectively, and is the cause of a potentially fatal disease.

Mutations can also result from the addition or subtraction of one or more pairs of bases, changing the gene by garbling its message. This type of mutation, which results generally from exposure to radiation, is more serious than the previous one. It can cause the incorporation of a sequence of incorrect amino acids in the protein, probably making the protein nonfunctional. In the case of albinism, the mutation does not make a difference in whether a person lives or dies. But if the mutation causes a disease such as diabetes, a person can die or have his or her life span severely reduced.

Substitution of one amino acid for another in a protein is the most common type of mutation and is responsible for most of the diversity among organisms, including humans. Let us go back to hemoglobin. Hemoglobin A and S are only two of the many forms of that protein. They happen to be the most common ones and for this reason were the first to be identified. And because hemoglobin S is responsible for a dreadful disease, hemoglobins A and S were both thoroughly studied. But since then, we have identified three hundred rare variants of hemoglobin and discovered that many other proteins have also a large number of variants. During the last twenty years, biologists have studied biochemical variation in many different organisms, including humans. Though we do not know how many different kinds of enzymes and other proteins are present in the human body, we can be sure that 10,000 of them is a conservative estimate. If we take this number and calculate the probability that two individuals randomly chosen from any community have the same proteins, we will find it very close to zero.[5] Geneticists would agree with Richard Lewontin that except for two identical twins, we are biologically unique, but, more interestingly, no human being ever was or ever will be identical to us.

Knowledge about enzymes and other proteins and blood types has permitted geneticists to answer the question of how much human diversity can be attributed to individual differences and how much to group differences. Using sophisticated statistical methods, they were able to estimate how much human diversity is contained within a given population and how much there is between large populations, such as nations and major "races" (as classified the old typological way). It seems that only a very small amount of genetic variation is explained by race. Richard Lewontin estimates it to be only six percent, but believes it to be even less.[6] He concludes that "the taxonomic division of human species into races places a completely disproportionate emphasis on a very small fraction of the total human diversity." That scientists as well as nonscientists nevertheless continue to emphasize these genetically minor differences and find new "scientific" justifications for doing so is an indication of the power of socioeconomically based ideology over the supposed objectivity of knowledge.

Notes for Chapter Twenty-One

1. Richard Lewontin, *Human Diversity* (New York: Scientific American Library, 1982).

2. We are giving here only a simple summary of the structure of DNA. The reader interested in more details can get them in any book of genetics.

3. This fact has permitted us to develop a technique that has helped criminologists track down murderers, rapists, and muggers. Virtually foolproof identification of any person is possible through a powerful laboratory test that detects so-called genetic fingerprintings in samples of blood, semen and hair roots. People have become well aware of the importance of DNA in a murder trial since the O. J. Simpson case.

4. Galileo Galilei marveled at the invention of the alphabet. In. *The Dialogues Concerning the Two Chief World Systems,* he wrote: "But surpassing all stupendous inventions, what sublimity of mind was his who dreamed of finding means to communicate his deepest thoughts to any other person, though distant by mighty intervals of place and time! Of talking with those who are in India; of speaking to those who are not yet born and will not be born for a thousand or ten thousand years; and with what facility, by the different arrangement of twenty six characters upon a page!" Trans. Stillman Drake. [This number should be 24 since there are 24 letters in the Italian alphabet, not 26 as in the English alphabet.]

5. Richard C. Lewontin. *Genetic Basis of Evolutionary Change* (New York: Columbia University Press, 1974).

6. Ibid.

Chapter 22

EXCEPT FOR A VERY FEW OF US, WE ARE ALL COLORED

People of browner complexions simply have more melanin in their skin. . . . It is not an all-or-nothing difference; it is a difference in proportion.

Ruth Benedict[1]

Skin color has been and is still used to physically describe people. It is easy to understand why. Skin color is one of the most striking human traits. All other distinctions fade before this one. A very dark-skinned or a very light-skinned person stands out among people who have skins of different colors. And that distinction remains, for no amount of sun, no skin treatment, no drug, can fundamentally and permanently change someone's skin color. Not only is skin color the most visible and striking human trait, it is also one whose social significance has been tremendous throughout our history.

People who happen to have more skin color than others have for centuries suffered widespread exclusion, exploitation, segregation, humiliation, and even death at the hands of those who happened to have far less skin color than they did. In the past, this single trait has served to identify the free and the enslaved, and to differentiate between the haves and the have-nots, those who could vote and those who could not, and between those who could give testimony in courts and those who could not. Today, there is still an unequal distribution of power and wealth between those called "whites" and those called "colored".

Today, when we turn on the radio, watch television, read the newspapers we are constantly reminded that there are "black" and "white" people, who live in black and white neighborhoods, go to black and white churches and in many cases to black and white schools and colleges. The words black and white are social terms, not biological terms. They are used to denote two ethnic groups which are misnamed because the skin color of each group varies from light to dark and because so-called "white" people are colored. For as we shall see in the next chapters, differences in skin color among us depend mostly on the amount of a pigment, called melanin, present in our skin. Except for albinos, all of us have some melanin in our skin, though some populations have, in general, more pigment than others. The amount of melanin in our skin varies tremendously from individual to individual, regardless of the group into which we happen to be classified. It even varies widely among members of the same family.

The true "whites" are the albinos. Their skin is not simply very light, but milk-white, their hair is whitish-yellow (straw-colored), and their eyes look pinkish, all because they lack the pigment melanin. Albinism occurs among all human groups, but at different frequencies. It occurs most frequently among the San Bias Indians of central Panama, where as many as seven infants per thousand births are reported to be albinos. The world wide frequency of albinism is about one in 20,000. Because of their lack of pigmentation, albinos have poor vision and are extremely prone to skin cancer.

Allusions to albinism date from antiquity, but the actual term albino, from the Latin *albinus* or white, was coined by the seventeenth century Portuguese explorer Balthazar Tellez, who sighted certain "white Negroes" on the west coast of Africa. Columbus claimed to have encountered such people near Trinidad at the time of his fourth voyage to America. Albinism is hereditary and occurs rather frequently within some families. It occurs not only in humans but throughout the animal kingdom, from reptiles and insects to fish, birds, and mammals. It is very common in mice, rats, rabbits and squirrels.

Though we all have melanin in our skin, the fact remains that skin color had assumed and still assumes a fantastic importance in our self-image and the conduct of our affairs because of myths about skin color. The most damaging of these myths, which persist in the minds of too many, is that "white" supremacy is due to biological superiority. In other words, the success of white people is attributed to the fact that they are light-skinned. On the other hand, it is assumed that there is something inherent in the biology of dark-skinned people that prevents them from being successful. In order to demean them, it was necessary to link differences in skin color with other characteristics—physical, behavioral, and intellectual. We have seen in chapter fourteen that there have been periodic attempts to link differences in skin color with differences in intelligence. These attempts have all failed. Yet, it was believed by many in the past—and it is still believed by some today—that the less colored your skin is, the smarter you are. Recently some black leaders have argued that the opposite is true. However, their position is no better than the opposite held by most light-skinned people in the past and by too many of them today. The amount of melanin has nothing to do with physical abilities, behavioral or mental abilities.

In chapter six I have emphasized that skin color is just that: skin color. It is a biological trait which, like height or any other physical trait, has nothing to do with the way the person thinks or acts.[2] Skin color is not acquired or possessed by leading a good or bad life, but is largely determined by our genes at conception.

The next three chapters deal with the biology of skin color, a subject of extreme importance, but rarely taught anywhere. This is unfortunate because we would learn that the differences in skin pigmentation that play such an important role socially are due to relatively small differences in the size and shape of pigment granules that are present in the skin of each of us.

Notes for Chapter Twenty-Two

1. Ruth Benedict, *Race, Science and Politics* (New York: The Viking Press, 1945).

2. Some people have even suggested that melanin "is responsible for a bonding with narcotics that makes people of African descent more vulnerable to drug addiction." *Lansing State Journal*, 28 March 1989, editorial page.

Chapter 23

CAN WE CHANGE OUR SKIN COLOR?

If a white man became a Negro in the Deep South, what adjustments would he have to make? What is it like to experience discrimination based on skin color, something over which one has no control.

John Howard Griffin[1]

William James[2] said once that there is no way to explain the experience of being in love to someone who has not had the experience.

In a similar way "black" people generally tell their "white" friends that no matter how they try there is no way for them to know how a black person feels when discriminated against since they are not themselves blacks. But we know that at least two people did share this experience simply by changing the color of their skin and for a few weeks successfully "passing" as "Negroes." One was John Howard Griffin in 1960 and the other was Grace Halsell in 1970.

John Howard Griffin related his experiences in Louisiana, Mississippi, Alabama and Georgia. There, as a "temporary" black man, he was subjected to racist insults, squalor, violence, antagonism, and hopelessness. He came back to his Texas home in Fort Worth to write his book, *Black Like Me*. Grace Halsell also wrote a book about her experience in Harlem and Mississippi as a "temporary" black woman. In her book, *Soul Sister*,[3] she described how racism permeated every situation she encountered.

Both Griffin and Halsell changed the color of their skin, but nothing else, not the shape of their nose or their hair. Yet, the change of skin color from light to dark was enough by itself to make the people who saw them treat them as members of the "Negro race." As such, they were subjected to racism, being constantly reminded of their inferior status, having to bypass available restrooms and eating facilities to find those that were specified for them, and being called disparaging names which referred to them as members of a subspecies well known to be lazy, dishonest, and stupid. But, in fact, Griffin and Halsell were exactly the same people in all their characteristics as they had been before their skin color was changed. How did John Howard Griffin and Grace Halsell temporarily change their skin color from white to black? Griffin told us that he took a medication orally, and exposed himself to ultraviolet rays. But he never told us which drug he took. Reading Halsell's book, however, we learn not only that she took psoralen, a photosensitizing drug that stimulates melanin production when coupled with regular and prolonged exposure to sunlight, but also that Griffin used the same drug. Both became dark in a few days. What did this drug do? It tremendously increased the amount of melanin produced in their skin. To understand how this is possible, we need to know where and how melanin is formed, that is, to know something about the biochemistry of melanin and the structure of the skin.

Melanin is present in very diverse groups of animals. For example, it is the black substance ejected by a squid when it is trying to get away from a predator. Melanin is a complex biological substance that is formed from the amino acid tyrosine. The formation takes place in a series of steps in each of which intermediate compounds are formed. Tyrosine is first transformed to a diphenol, a colorless product, and then to a quinone, which is colored. An enzyme, called tyrosinase, acts as a catalyst in both of these transformations. Differences in skin color among us are due in great part to the differences in the activity of this enzyme. In albinos, tyrosinase does not function, and melanin cannot be formed. In individuals whose skin is light, tyrosinase exists in a partially inactive form during the winter months. But throughout the rest of the year it becomes active, because the sunlight gradually destroys the substances that inhibit its activity. Hence, the formation of a suntan. In those individuals with dark skin, tyrosinase is active throughout the year.

Melanin is a very stable pigment, highly insoluble in common solvents.[4] It is so resistant to decomposition that it has been found not only in the skin of mummies and of an extinct mammoth, but also in that of a dinosaur, a 150-million-year-old ichthyosaurus. We know that melanin is formed in our skin. But where exactly and how? Most of our knowledge about the structure and color of human skin dates from the last forty years. It started when medical scientists became interested in the disorders of skin pigmentation. Thanks to their biochemical and microscopical studies, we now know a lot about this subject.

Although studying skin structure with the electron microscope has cleared up many problems, there is still much to learn.

Human skin is structurally divided into three distinct layers: The top layer is called the epidermis, the thick middle layer is the dermis, and the lowest layer is the subcutaneous fatty tissue (fig. 1). The epidermis evolves from a dense population of actively dividing cells.

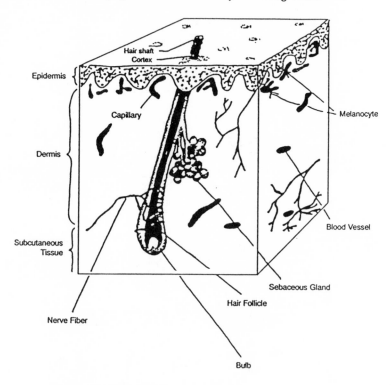

Figure 1. The skin and its most important parts

The surface of the epidermis, a thin layer of remnants of dead cells, is called the stratum corneum. Its primary function is to maintain a tough protective barrier over the entire surface of the body. The dermis, the thickest part of the skin, is composed of dense tissue in which there are embedded fine blood vessels, various nerves (especially those sensitive to touch), the smooth muscles that raise the hair when contracted, and a variety of specialized glands. The deepest layer, the subcutaneous fatty tissue, is characterized by closely packed cells that contain considerable fat. This is the layer which provides thermal insulation, reducing loss of body heat.

Our differences in skin color are primarily due to the amount of melanin that is present in the epidermal layer of our skin.[5] This pigment is synthesized in specialized cells called melanocytes,[6] located between the dermis and the epidermis (fig. 2A). These cells, which have long tentaclelike projections, inject granules of melanin into the surrounding epidermal cells (fig. 2B). There the pigment seems to form a protective awning over each cell nucleus on the side toward the skin surface.

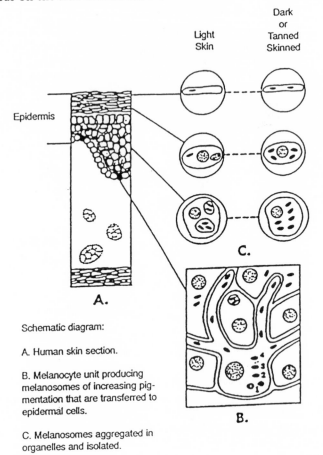

A.

Schematic diagram:

A. Human skin section.

B. Melanocyte unit producing melanosomes of increasing pigmentation that are transferred to epidermal cells.

C. Melanosomes aggregated in organelles and isolated.

Figure 2 (A, B, C).

Source: See note 7

It is believed that the primary function of melanin is to shield the cell nucleus by absorbing the ultraviolet rays of the sun, which can damage the DNA and cause mutation in the genes contained in the chromosomes that are in the nucleus of the epidermal cells. Many of the skin cancers are the consequences of these mutations. Within the melanocytes there are even smaller structures, called melanosomes, in which melanin is produced. As the melanosomes develop, they go through various stages.

As shown in figure 3 in stage one, the melanosome is spherical and has no recognizable internal structure. No melanin is present in it at this stage of development. In stage two, the melanosome is now oval and has distinct internal structure made up of more or less parallel layers. A small amount of melanin is evident in this stage. In stage three, the melanosome contains a larger amount of melanin. At this stage the layers visible in stage two are largely hidden by this melanin. Finally, in stage four, the melanosome has the maximum amount of melanin that it can have.

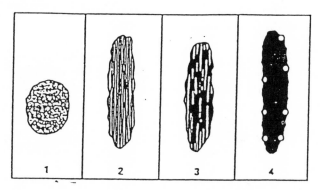

Figure 3. Four Stages in Melanosome Development

The melanocytes of the darker skin of Asians, American Indians and East Indians generally contain moderately melanized stage three and four melanosomes. On the other hand, the melanocytes of very dark-skinned people, such as Africans and Australian aborigines, are usually filled with stage four melanosomes with only a few stage two and three melanosomes. Before irradiation with ultraviolet light, the melanocytes of individuals with very light skin have very few melanosomes in any stage of development and those of light-skinned individuals contain a large number of stage two and three melanosomes and may contain a few four stage melanosomes. Once mature, the melanosomes travel to the tips of the tentaclelike projections of the melanocytes and then are engulfed by the *keratinocytes*, the cells that make up the stratum corneum layer, which come in contact with them. The melanosomes then migrate with the epidermal cells toward the surface of the skin. We do not know what it is that moves the melanosomes through the epidermis. This is common in science: We can describe some phenomenon, but we often do not know what causes the phenomenon to occur. Though there is a tendency for more deeply pigmented skin to contain larger melanosomes, the main distinctive feature that distinguishes the skin of a dark-skinned individual from a light-skinned individual is how the melanosomes are arranged in the skin cells. Melanosomes can occur as single units if they are large. If they are small, they tend to group together, three to five, in a transparent structure like a plastic bag (see fig. 2C).

In dark-pigmented Africans and Australians the melanosomes are mostly single, whereas the skin of Europeans and Asians, when not exposed to sunlight, contains melanosomes in groups. But, we should emphasize that both single and aggregated melanosomes are observed in all types of skin. Children of European- African parents have both types of melanosomes.

As the melanosomes travel up through the epidermis, they tend to break into amorphous melanin particles. This is particularly true of the small melanosomes which are grouped together. The disintegration of melanosomes is due to the action of very strong enzymes and occurs within the cells of the upper epidermis.

Regardless of how dark or how light our skin appears, we have the same skin structure, we produce the same type of melanin, and we have the same number of melanocytes per unit of skin area.[8] What causes the striking differences in skin color that we see is how much melanin the melanocytes produced and how large and how many melanosomes are present in the surface layer of our skin. In other words, the variation in skin color does not occur at the structural level but at the functional level. It is not an all-or-nothing matter, but a question of degree. But one thing remains to be explained. The differences in skin color are extremely striking to the naked eye and yet these differences tend to be reduced under the microscope. To understand this paradox, let us take a page of a newspaper with pictures. To the naked eye, the dark areas in the pictures appear fully black. However, when seen under magnification, a hand lens for example, the dark areas now are seen to be formed of many dots with spaces in between. The larger the spots are and the closer they are to each other, the blacker the image appears to the naked eye. The same is true for the skin. The dots are the melanosomes. The larger they are and the closer they are to one another, the darker the skin is. The fewer the melanosomes the lighter the skin appears because there is a lot of space between the melanosomes. Another reason why naked-eye observations differ from microscopical observations is that in the first case, light is reflected from the skin, and in the second case, the light is transmitted through a thin slice of the skin. The picture we get is different. To understand this, let us take a color slide and a color print of the same object. To see the object in the slide we need to transmit light through it. However, we see the object in the color print by reflected light. The picture that we get from the slide reveals far more detail and has a far larger range of brightness than the picture from the color print. The same is true if we photograph the painted windows in a church from the inside and from the outside. We get far more details in the first case than in the second case.

The most surprising discovery that came from the microscopical observations of human skin has to do with tanning, which we will discuss in the next chapter. We will also explain further how John Howard Griffin and Grace Halsell were able to temporarily change their skin color from light to extreme dark and pass for Negroes.

Notes for Chapter Twenty-Three

1. John Howard Griffin, *Black like Me* (Boston: Houghton Mifflin Co., 1960).

2. William James wrote: "One must have been in love one's self to understand the lover's frame of mind."

3. Grace Halsell, *Soul Sister* (Cleveland: The World Publishing Co., 1970).

4. Recently, it has been shown that melanin can be dissolved. Giuseppe Prota, *Melanins and Melanogenesis* (New York: Academic Press, 1992).

5. In individual cases, the appearance of skin pigmentation is also influenced by the structure of the skin: its thickness, translucency and oiliness and the whole process may be intensified or diminished by hormonal action. People also vary in the amount of moles and freckles. Moles are clusters of melanocytes which begin to appear on the body in the third year of life and slowly increase in number with age. Genetic and hormonal factors control the size and number of moles in the individuals. In rare instances, excessive concentrations of melanocytes appear in some parts of the skin before birth, producing large moles. Freckles are caused not by an excess of melanocytes, but simply by a higher concentration of melanin. In individuals with the appropriate genetic makeup, repeated exposure to sunlight causes freckles to appear beginning in about the sixth year of life.

6. The melanocytes originate as part of the embryonic nervous system from the ridge of the spinal cord and migrate along its sides to the skin. This occurs from about the eleventh week until the end of the four months of embryonic life. The fact that melanocytes are originally nerve cells explains why they have tentaclelike projections and why skin pigmentation is influenced so much by hormones.

7. Quevedo, et al. "The Role of Light in Human Skin Variation," *American Journal of Physical Anthropology* 43 (1975): 393–408.

8. However, the melanocytes of red-headed Europeans are small with few dentrites (the technical term for the tentacle projections), a possible reason why these people are more susceptible to sunburn.

Chapter 24

NOTHING UNDER THE SUN IS JUST
BLACK OR WHITE

Now it is fashionable to be dark, but only if you are really white, that is tan is beautiful—if you keep it this side of black.

Grace Halsell, in *Soul Sister*

If we bring up a dark-skinned child born in Africa in northern France, or if we bring up a light-skinned child, born in France, in equatorial Africa, they will remain, respectively, dark and light. Something fundamental, therefore, must exist in their genetic makeup that determines their basic pigmentation. Scientists call this basic pigmentation *constitutive skin color*. If our degree of pigmentation varies with our genetic constitution,[1] it also varies with the environment. Some of us are lightly pigmented a part of the year and may become more pigmented the rest of the year, depending upon how much our skin is exposed to the sun. We call this increase in skin pigmentation above the level of natural pigmentation a tan. Scientists have another phrase for it: inducible skin color. Today, and particularly in the western world, the pursuit of a tan has become a passion, and there are people who will spend hours sunbathing or in tanning salons. The achievement of a bronzed appearance is believed to signify health and beauty, as long as you "keep it this side of black." The biological side of the coin is that too long and too frequent exposures to the sun (or artificial sunlight) can have very harmful effects in light-skinned, and not so light-skinned individuals. This is particularly true for those individuals who have light-colored eyes, red or blond hair and freckled skin, for many of them do not ever tan. If their skin is exposed to the sun, it reddens, burns easily, blisters and peels. These people are not only susceptible to sunburn but also to skin cancer. Light-skinned individuals who can tan are less affected by solar rays, but should still be cautious about sunbathing, for lifelong exposure to solar radiation has been associated not only skin cancer but also with eye cataracts. And if the threats of cancer or cataracts are not deterrent enough, sunbathers should worry about wrinkles, for sun radiation damages the protein building blocks of skin. The skin will sag, become drier, nodular and pebbly—all changes that we associate with aging. There are many cases of fifty year-old individuals, who, after a long life of sunbathing, look seventy. Skin cancer is increasing not only among the United States population which, since World War II, has adopted the sun-worshipping culture, but also among the light-skinned descendants of the settlers of Australia and even among the descendants of Japanese immigrants in the Hawaiian Islands. Both these islands and Australia are close to the tropics where sunlight is the most intense. The most fortunate individuals are those who are constitutively dark-skinned or who tan very fast.

It is important to note that these skin cancer studies could be improved if researchers were to abandon entirely the concept of race in tabulating their data.[2] The inclusion of something so artificial as race masks the true meaning of research results, since it is logical to assume that susceptibility to skin cancer is inversely related to the amount of melanin in a person's skin regardless of his or her "race." Researchers have ignored evidence that there is tremendous variability in tanning ability among the light-skinned individuals. If they were to classify people according to their skin color in the summer, their data might be more meaningful than when race enters into the picture. After all, it is reasonable to assume that someone who does not tan easily is more susceptible to skin cancer than someone who tans rapidly regardless of whether he or she is considered black, white, or Native American. The component of sunlight that plays a role in skin burning and the formation of skin cancer is also the one that induces tanning. Herein lies a paradox for those of us who are light-skinned: If we protect ourselves from sun radiation, we do not get burned, but we do not get tanned either. If we expose ourselves to the sun, we often get burned before getting that tan which protects us from burning. The culprit is known as ultraviolet light, which is part of the electromagnetic spectrum and lies between the visible light and X-ray regions (fig. 1).

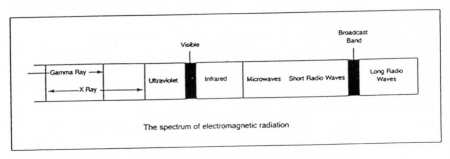

The spectrum of electromagnetic radiation

Figure 1. The spectrum of electromagnetic radiation

Inside the sun, thermonuclear reactions convert millions of tons of hydrogen units into millions of tons of helium units every second. During this process, large quantities of matter are destroyed and released from the sun's surface in the form of energy. This radiant energy spreads in every direction at the speed of light and reaches the outer limits of the earth's atmosphere. The components of the sun's radiation may be characterized by wavelength, or energy. In ascending order the solar spectrum includes (1) short-wave, high-energy X-rays; (2) ultraviolet rays; (3) visible light; (4) infrared rays; and (5) long-wave, low-energy radio waves. In general, about one half of the energy from the sun that reaches the earth's surface is in the form of infrared radiation, and the remaining half consists mostly of visible and ultraviolet light. The other forms of variation, such as X-rays, are screened out by the atmosphere. The same is true of the shortest wavelength (highest energy) ultraviolet radiation. In the process of passing through the atmosphere, ultraviolet, visible, and infrared radiation are absorbed, scattered without loss of energy, and reflected in every direction, depending on the specific wavelength involved and the degree of atmospheric turgidity and cloudiness. The visible portion is least affected by atmospheric conditions, while water vapor is especially influential in reducing the amounts of infrared rays striking the earth's surface. Infrared is the part of the spectrum that causes a rise in skin temperature and provides the warm feeling one gets while in the sunlight. Ultraviolet radiation has been divided into three segments: UV-A, UV-B and UV-C. UV-C will not be mentioned further because, having wavelength of less than 280 nanometers, it does not reach the earth, being screened out by the ozone layer. On the other hand, the ultraviolet portion is absorbed by ozone but is relatively unaffected by water vapor, a factor which explains why one can become sunburned even on an overcast summer day. If the ozone layer is indeed in the process of depletion, as some ecologists have suggested, more ultraviolet radiation will reach the earth and cause trouble, contributing, among other things, to a rise in skin cancer and deaths. Contrary to the infrared radiation that penetrates the skin deeply, very little ultraviolet radiation enters the lower layer of the epidermis. Dark pigmented skin is penetrated much less than fair skin.

With exposure to ultraviolet radiation, the skin of every one, except those whose skin is very dark, becomes tanned by two distinct processes: immediate tanning and delayed tanning. Immediate tanning is induced by the longer wavelength, UV-A; it becomes most prominent within one hour of exposure and almost completely disappears within four hours. This rapid pigmentation involves an increase of melanin in the melanosomes and a rapid transfer of these melanosomes from the melanocytes that produced them to the surface of the skin, but there is no increase in the number of melanosomes. On the other hand, delayed tanning, which is induced by the shorter wavelength ultraviolet, called UV-B, develops four to five days after exposure to the sun radiation. It involves new production, transfer, distribution and degradation of melanosomes. Pigmentation appears slowly over a period of days. Inasmuch as new melanosomes are synthesized, an increase tyrosinase reaction can be clearly demonstrated. Exposure to ultraviolet radiation also causes the epidermis to thicken, which increases tolerance to subsequent irradiation. This response results from increased cell multiplication in the upper and lower epidermis. Little is known how this process occurs. The skin remains tanned for several weeks and offers considerable protection against further tissue damage by sunlight. Eventually the pigmented cells slough off and the tan slowly fades.

I have said in the preceding chapter that the most striking difference under the microscope differentiates people as to skin pigmentation was that melanosomes can be single or form complexes. In the skin of dark-skinned individuals the melanosomes mostly occur as single units surrounded by a limiting membrane, and in the unexposed skin of light-skinned individuals, the melanosomes are aggregated, packaged three to five particles within a limiting membrane. But, all these differences tend to disappear after exposure to ultraviolet radiation as electron microscopical studies have shown. Tanned skin contains melanosomes in all stages of development [3,4] and many of them are found singly the way we find them in the skin of dark-skinned individuals (figure 2). Therefore, we should expect that the tanned skin of light-skinned individuals would be indistinguishable from the skin of dark-skinned individuals. Though this may be true under the microscope, it is not true to the naked eye. The tanned skin of a light-skinned individual never reaches the intensity of color of the skin of someone who is constitutively extremely dark-skinned, possibly because it contains fewer melanosomes.

In the case of John Griffin and Grace Halsell, however, their skins were very black because they ingested the drug psolaren, which, in concert with ultraviolet radiation that they received, induced the melanocytes in their skin to produce larger melanosomes and in greater numbers than with ultraviolet radiation alone, permitting them to look like, and pass for, "Negroes." The mechanism by which administration of psolaren and subsequent exposure to ultraviolet radiation causes this extreme skin pigmentation remains mostly unknown. Nevertheless, it is possible to change one's skin color from light to dark. But so far, no one has been able to change one's skin color from dark to light.

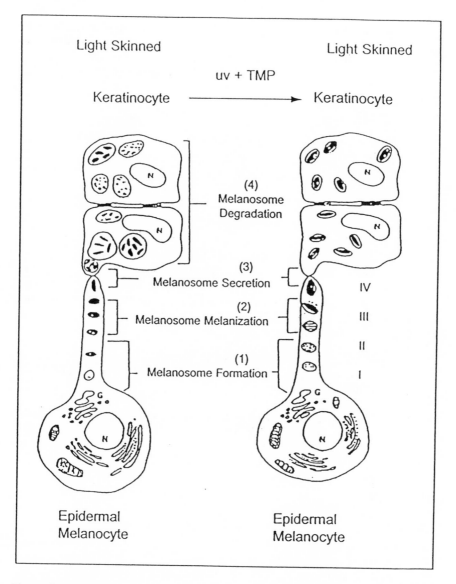

Figure 2. Facultative and constitutive melanin pigmentation in the upper arm of an American of European ancestry. The untanned skin exhibits constitutive pigmentation, whereas the tanned skin displays facultative pigmentation. Within the untanned skin, melanosomes mainly form melanosome complexes. Within the tanned skin, melanosomes treated with psoralen are larger and generally tend to be arranged singly and are indistinguishable from those of the constitutive pigmentation of dark-skinned Africans or Australians.

Source : see note 2

Notes to Chapter Twenty-Four

1. We have not been created equal as to our reaction to sunlight, and dermatologists distinguish six types of skin according to their sensitivity to sun radiation and their pigmentary responses. They are indicated below according to skin type, skin sensitivity, and pigmentary response:

    1. *Very sensitive, always burns. Tans little or not all easily even with repeated exposures.*

    2. *Very sensitive, always burns. Tans minimally with repeated exposures.*

    3. *Sensitive, burns moderately. Tans gradually (light browns).*

    4. *Moderately sensitive, burns. Tans easily (moderately brown).*

    5. *Minimally sensitive, burns rarely. Tans profusely (dark brown).*

    6. *Insensitive, never burns. Extreme, deeply pigmented (black) tan.*

    After M. A. Pathak's "Epidermal Melanin Pigmentation Stimulated by Ultraviolet Radiation and Psoralens." In *Brown Melanoderma*, T. Fitzpatrick, M. Wick, and K. Toda, eds. (University of Tokyo Press, 1986).

2. Ki Yoshi Toda et. al., "Alterations of Racial Differences in Melanosome Distribution in Human Epidermis After Exposure to Ultraviolet Light." *Nature* 236 (1972): 143-145.

3. W.C. Quevedo et. al., *Light and Skin Color, In Sunlight and Man*, Pathak et. al., eds. (University of Tokyo Press, 1972.)

4. Pathak, "Epidermal Melanin Pigmentation."

Chapter 25

GENES AND SKIN COLOR: THE MORE THE MERRIER

The number of genes determining skin color inheritance in humans cannot be determined until detailed family studies have been undertaken.

Pamela Byard[1]

Skin color is undoubtedly inherited, but misconceptions about it are many and have been repeated in literature. For instance, in one of Conan Doyle's Sherlock Holmes adventures, a little girl wears a yellow mask. The reason for this strange behavior is that the mother is afraid she would lose the love of her husband if he knew that she had a child from a former husband, an African American, who had died three years earlier. Showing the portrait of her deceased husband to the astonished Holmes, Watson and her second husband, the woman said:

> That is John Hector of Atlanta and a nobler man never walked the earth. I cut myself off from my race in order to wed him but never once while he lived did I for an instant regret it. It was my misfortune that our only child took after his people rather than mine. It is often so in such matches, and little Lucy is darker than her father was. But dark or fair, she is my own little girly and her mother's pet.[2]

There is a problem with this story. It is impossible for a child who has one very light-skinned parent, as Lucy did (her mother), and one dark-skinned parent (her father), to be even as dark as the darker parent, let alone darker than that parent. It can be stated with certainty that no authenticated instance of such a birth has ever been reported. Nor has there been any well-documented case in which dark-skinned children were born to very light-skinned parents. On the contrary, in the few instances in which this was alleged to have happened, investigation revealed that the reported instance was based on hearsay and not observed fact, or that it was the result of concealed illegitimacy involving a darkly-pigmented parent.

We can make allowance for Conan Doyle's mistaken idea. After all, in his time, the science of genetics had not been born, and no one could have explained to him why little Lucy could not be darker than her father. For approximately eighty years we have tried to explain skin-color differences by assuming that from two to five pairs of genes were involved, that very light-skinned individuals have no genes for pigmentation, and that very dark-skinned individuals have all of them. It was further assumed that there was no dominance and recessiveness in the genes for skin pigmentation, and that the effects of the genes involved were additive, i.e., that one individual who has two genes for pigmentation would be twice as dark as one who has only one. It follows from these assumptions that no genes for color are in hiding. "White," i.e., very light-skinned persons, cannot transmit to their children any gene for pigmentation that they do not have. The only genes for pigmentation that these children have come from the other parent if he or she happens to have a dark complexion. Therefore, children cannot have more genes for pigmentation than their darker parent and cannot be darker than him or her.

The first studies of the genetics of skin color date from 1913, when Charles Davenport and Gertrude Davenport.[3] became interested in the inheritance of differences in skin color. They carried out their investigations in Jamaica and Bermuda, where marriages among "blacks and whites" are fairly common, since they were not forbidden as they were in many places in the United States. There, the ancestry of the parents could be traced fairly clearly. The reader should note that, since all of us have some color in our skin, the problem that the Davenports attacked was different from the problem of inheritance of skin color in general. They were dealing with the inheritance of the differences in skin color between two human groups.

The investigators were able to find six families in which one of the grandparents was light-skinned and the other dark, and which produced thirty-two grandchildren. They measured skin color of members by means of a "color top." Colored paper disks were overlapped in such a way that varying proportions of each color were exposed. When the top was spun, the colors seemed to blend together. By varying the proportionate amounts of the colors black, yellow, red and white, the investigators could match the skin color of persons who were studied. Data were expressed by the percentage of black on the color top. For example five percent meant that five percent of the black disk on the color top was exposed in the matching blend.

The thirty-two offspring were classified into five different skin color categories, ranging between the extremes of the grandparents, with some lighter and some darker than their parents. The choice of five categories was made on the basis that five peaks of frequencies of pigmentation types seemed discernible. The investigators concluded from their data that only two pairs of genes were involved. But their analysis was not adequate for two reasons: (1) Their sample was too small, and (2) there was a considerable range of pigmentation within any given category. This meant that more than two pairs of genes had to be involved in determining the differences in skin color among these thirty-two individuals. The number of individuals being small, it was a safe bet that the range of colors was not completely represented in the sample, and that the number of genes determining the differences in skin color among the group in particular and human beings in general is far greater than two pairs. In 1949, Ruggles Gates,[4] studying families of black American populations, concluded that three pairs of genes of unequal effect were involved. A similar conclusion was reached by Curt Stern in 1953[5], and later in 1970[6], when he reexamined the 1953 data. To estimate the number of genes involved in skin color, Stern used an original method. He set up a series of alternative genetic models, then compared their consequences with empirical data. Basically he compared the observed frequency distribution for skin pigmentation in black American populations with expected frequencies based upon models using two, four, six, ten and twenty pairs of genes. He concluded that models involving three to four pairs of genes fit the observed distribution best.

But all of these early studies were not very good because they lacked an adequate method of measuring skin color. The color top or colored plate methods were not satisfactory because they depended too heavily on the subjectivity of visual matching—the eyeball method. For example, since an exact match of the observed skin was not always found, two observers might place the same individual into different classes.

At the present time, a more objective and consistent evaluation of skin color can be obtained by the use of a spectrophotometer.[7] This casts a beam of light on a skin area of the person being examined. This apparatus measures how much of such a beam of light of a particular color is reflected by the skin. The more melanin there is in the skin, the less the amount of light that is reflected. The reflected light from this lighted area enters a sensitive photoelectric cell. This cell activates a recording device which indicates the percentage of reflectance of each wavelength of light the cell receives. A graph resulting from the observations shows the percentage of reflectance at each wavelength. These spectrophotometric measurements of skin color provided evidence that there was a continuous variation in melanin amounts and suggested that many genes were involved in the inheritance of differences in skin color. Yet, geneticists persisted in believing that only a few pairs of genes, three to five, were sufficient to account for the differences in skin color, not only between Africans and Europeans but between Europeans and Vietnamese, Europeans and American Indians, and Europeans and Australian aborigines.

Their hypothesis was widely accepted for years and has been widely publicized in textbooks in human biology. However, in 1981, Stern's method of constructing models to determine the number of gene pairs in a hybrid population was criticized by Pamela Byard and Francis Lees.[8] They pointed out that there was a basic flaw in Stern's analysis, namely, that the number of genes that one determined by his method depended on the number of classes into which the observed values were divided. For example, if we divide the range of colors into three classes, blacks, browns, and whites, the least number of pairs of genes that can explain the data is one pair. If symbolically A1A1 determines black, A1A2 determines brown, and A2A2 determines white, then marriages between two brown-skinned people would give offspring of only three different colors: black, brown, and whites, in the proportion respectively of 1,2,1. If, on the other hand, we divide the same range of pigmentation into five classes, we can explain the data by assuming two pairs of genes. This is how the Davenports explained their data. Curt Stern, who criticized their work, fell into the same trap. When he arbitrarily divided the color range of human skin into eleven classes, he automatically had to find that the number of pairs of genes involved was five. If Stern had divided the range into more classes, he would have found that the same data would be explained by a greater number of pairs of genes.

The relationship between the number of pairs of genes and the number of classes, in which the range of skin color of children of two medium-skin-color parents is divided, is given below.

Number of pairs of genes	Number of Classes
1	3
2	5
3	7
4	9
5	11
6	13
n	$2n+1$

To understand better the flaw that is inherent in this approach in determining the number of genes of a trait, let us turn to something with which everyone is, or should be, familiar: the determination of course grades. Suppose a geneticist, who assumed that the ability to sucessfully pass exams is in great part genetic, wants to learn how many pairs of genes are involved in this process. He gives an exam to 5000 students and records their scores on a percentage basis. But, teachers have to give grades. They have to classify students in categories. Generally these categories are A, B, C, D and F. As soon as our teacher of genetics classifies his or her students into this number of categories, he or she falls into the trap of the early geneticists who researched the inheritance of skin color. Our researcher is automatically led to the conclusion that there are two and only two pairs of genes involved in the inheritance of the ability to pass exams. If the number of categories of grades were increased, let us say to 13 (A+, A, A-, B+, B, B-, C+, C, C-, D+, D, D-, F), our misguided geneticist would conclude that there are six pairs of genes involved, which is as erroneous as the two pairs he had found previously. This division into grade types is artificial, and so was the division into color types used by Davenport and Stern. Actually, there is, as every teacher knows, a continuous distribution of scores, which, when plotted on a graph, approaches a normal curve. This is true also for skin color data, in particular from reflectance spectrophotometry.

The only conclusion that we can draw from our studies of inheritance of skin color is that many genes are involved. This is not surprising since it is hard to believe that only a few genes would be responsible for the observed variations in the number, size, melanization, and dispersion of the melanosomes. On the other hand, differences in the degree of degradation of melanosomes that exist between dark- and light-skinned individuals may well serve to widen the variation in skin color and lead to exaggerated estimates of genetic differences.

Whatever the exact number of genes is that determines skin color, their number is very small compared to the tens of thousands that we have. It is unfortunate that the worth of a person has been judged, and is still judged, on the expression of a relative small number of genes, which have seemingly no other function than the determination of skin color.

Notes for Chapter 25

1. In a letter to the author dated October 27, 1992.

2. Arthur Conan Doyle, "The Yellow Face," in *The Complete Sherlock Holmes* (New York: Doubleday and Co., 1930).

3. Charles B. Davenport and Gertrude Davenport, *Skin Color in Negro and White Crosses*, Publication 188. (Carnegie Institute of Washington, 1913.)

4. R. R. Gates, *Pedigrees of Negro Families* (Philadelphia: Blackinston, 1949).

5. C. Stern, "Model Estimates of the Frequency of White, Near White Segregants in the American Negro," *Acta Genetica*. stat. med 4(1953): 281–298.

6. C. Stern, "Model Estimates of the Number of Gene Pairs Involved in Pigmentation Variability of the Negro-American," *Human Heredity* 20 (1970): 165–168.

7. For a detailed history, description, and discussion of the properties of reflectance spectrophotometry, see Ashley Robins, *Biological Perspectives on Human Pigmentation*. Cambridge University Press, 1991.

8. Pamela Byard and F. C. Lees, "Estimating the Number of Loci Determining Skin Color in a Hybrid Population," *Annals of Human Biology* 8 (1981): 49–58.

CONCLUDING THOUGHTS

"Yesterday's science is today's common sense and tomorrow's nonsense." For the concept of race . . . tomorrow is here.

Frank B. Livingstone[1].

Most people believe that human races exist and that their belief has been supported by scientific evidence. Nothing could be further from the truth. In spite of efforts, scientists have failed to demonstrate that humanity can be divided into races, i.e., groups of human beings that can be distinguished biologically. The reason for this failure is that humanity is so highly diverse that whatever trait is used to classify people into groups, there are always members of one group sharing this particular characteristic with members of several groups, rendering the classification system unworkable.

The biological reason for this is clear. If homogeneous groups or races, are to occur within any species, populations of that organism must have been sexually separated from one another for many, many generations. In this case, any difference that may have occurred within one group cannot spread to another. This sexual isolation has occurred within plant and animal species, but has never occurred with human beings. Geographical and social barriers have never been great enough to prevent members of one population from breeding with members of another. Therefore, any characteristic which may have arisen in one population at one time will be transferred later to other populations through mating.

Scientists have been unable to classify humanity into races using physical characteristics such as skin color, shape of nose or hair, eye color, brain size, etc. They also have been unable to use characteristics such as blood type or other genetic markers. Yet, many continue to believe in the existence of races, looking again and again for methods that would help them to find the elusive identifiers of "race." Why have scientists not abandoned the concept of race? Because most are captive of the same social and world view as nonscientists. For centuries, race has been an integral part of this outlook. It has been held that human beings are divided biologically into discrete and exclusive groups and that these groups are, by nature, unequal and that they can be ranked according to a superiority-inferiority scale. One consequence of this view has been the justification for the enslavement of those who are seen as physically different and therefore considered to be inferior. There is no doubt that the economic foundation of the modern world has been based on the preservation of these so-called racial distinctions. Hence, scientists raised in such a social environment have been led to search for biological bases for racial distinctions rather than to inquire as to whether or not races existed.

And although the belief in the existence of biological human races long remained unchallenged, about twenty-five years ago, a few bold anthropologists confronted the race dogma. The result is that today most anthropologists no longer think of human diversity in terms of race. Unfortunately, they failed to communicate their findings, that races were, and are, figments of our imagination, to the public at large. That is why so many of us still firmly believe that human beings are naturally divided into biologically distinct groups and why we treat each group as if it were a different species, rather than part of the same species.

Though we are all human beings, we are biologically unique. We have thousands and thousands of genes in each of our cells. Due to the processes of mutation, each of these genes can have different forms. The number of possible combinations of these different forms is infinite. Hence, the chances for an individual having the same combination of gene-forms as another individual are nil. This is true even for children of the same parents. This diversity is insured by the natural phenomenon of meiosis. Meiosis is generally unknown to most people, and its importance and significance are generally minimized in biological science courses taken by college students.

Another reason why each of us is different from everyone else is that we are the product not only of a unique set of genes, but also of a unique environment in which we grow and develop. The interaction of a unique set of genes and a unique environment assures our singleness, physical as well as mental. The fact that each of us is unique explains why biologists have discovered that diversity within a group is always far greater than diversity among groups, whatever the criterion used in the classification. The concept that each of us is a unique individual should be taught in biological science courses as well as social science courses. If it were, this would help in understanding why we should be treated as individuals and not as members of a specific, presumed biological group.

What are the consequences of abandoning race thinking? I can imagine some important ones in the realm of science as well as in society.

I have stressed in this book that race thinking has handicapped scientific research, in particular the field of medicine. The most striking example has to do with skin cancer. It is very likely that there is a clear relationship between the amount of melanin in someone's skin and his or her susceptibility to skin cancer. One would expect the number of skin cancer cases to be reported according to different skin types, from very light to very dark. However, the belief in the reality of races still dominates the thinking of medical researchers, and this is reflected in the statistical tables where such results are reported. In those tables data about blacks is separated from that about whites within each skin type. Doing this is illogical, since the question asked in this research is not whether the person is black or white, but what is the relationship between the amount of melanin in the skin and the individual's susceptibility to skin cancer. Given this, the individual should be classified only on the basis of the amount of melanin present in the skin and not also upon some assumed racial membership. The pooling of the data from both categories, black and white, might shed a better light on the inverse relationship between melanin content in the skin and susceptibility to this type of cancer.

Abandoning race thinking would help us to focus on the real causes of human diversity. Scientists have wasted too much time dreaming up possible reasons for the existence of the so-called racial traits. In spite of all of their efforts, they have been unable to find any. The closest they came to finding one was for skin color. However, even in this case, the evidence thus far is inconclusive and somewhat contradictory. One possible reason is that our differences in skin color are, like our differences in other characteristics, a question of degree, not of absoluteness.

Abandoning race thinking would also help us to focus on the real causes of why some individuals are unprepared to benefit from education instead of blaming their poor test performance on their racial background. All the evidence so far points out that the so much discussed differences in IQ between whites, African Americans, and Asian Americans, are not due to biological factors but to social and cultural ones.

But, the most important consequence of abandoning the idea that human races exist is the inspiring possibility that relations between human beings could be vastly improved. We would not be tempted as much as we are today to stereotype people; we would be more likely to respond to someone as an individual rather than as a member of a group.

I am not naive. I realize that this change in our attitudes will not happen overnight. If it is rather easy for scientists not to think any more in terms of race in their work, it is far harder for the public at large to do so in their daily life. Thirty years ago, hope was high that race thinking was on the decline, but today it is very much alive and on the rise, despite the best efforts of concerned individuals. Individual rights have been replaced by group rights with the dire consequence of a reverse trend for segregation which, according to Thurgood Marshall,[2] is the worst thing that ever happened to the United States. People have forgotten the idea stressed by Martin Luther King—that in a multiethnic society, no group can make it alone. To quote King:

> Our cultural patterns are an amalgam of black and white. Our destinies are tied together. There is no separate black path to power and fulfillment that does not have to intersect with white roots. Somewhere along the way the two must join together, black and white together, we shall overcome, and I still believe it.

Martin Luther King's famous speech, "I Have a Dream," given August 28, 1963 on the steps of the Lincoln Memorial was about integration. This is what he said:

> I have a dream that one day on the red hills of Georgia the sons of former slaves and the sons of slave owners will be able to sit down together at the table of brotherhood . . . I have a dream that one day the state of Alabama . . . will be transformed into a situation where little black boys and black girls will be able to join hands with little white boys and white girls and walk together as sisters and brothers . . . With this faith we will be able to work together, to struggle together, to go to jail together, to stand for freedom together, knowing that we will be free one day.

My dream is the same as that of Martin Luther King. My vision of the future is a colorblind society. I know that I will never see it. I doubt that my children will see it in their lifetimes; but I hope that my grandchildren will see it. If they do, they may remember the very small part that their grandfather played in helping to destroy the myth of human races.

Notes to Concluding Thoughts

1. Frank B. Livingstone, "On the Non-Existence of Human Races," in *The Concept of Race,* ed. Ashley Montagu (New York: The Free Press, 1962), p. 59.

2. Carl Rowan, *Dream Makers and Breakers: The World of Thurgood Marshall* (New York: Little, Brown and Co., 1992), p. 159.

SELECTED BIBLIOGRAPHY

Banton, Michael and Johnathan Hardwood. *The Race Concept.* New York: Prager Publishers, 1975.

Barkan, Elazar. *The Retreat of Scientific Racism. Changing Concepts of Race in Britain and the United States between World Wars.* Cambridge: Cambridge University Press, 1992.

Benedict, Ruth and Gene Weltfish. *Race Science and Politics.* New York: Viking Press, 1945.

Block, N. J. and G. Dworkin. *The IQ controversy.* New York: Pantheon Books, 1976.

Boyd, W.C. *Genetics and The Races of Man.* Boston: Little, Brown and Co., 1958.

Carnoy, Martin. *Faded Dreams. The Politics and Economics of Race in America.* Cambridge: Cambridge University Press, 1994.

Daedalus. "Color and Race," *Journal of the Academy of Arts and Sciences* 96, no.2 (spring 1967).

———. "Slavery, Colonialism, and Racism." *Journal of the Academy of Arts and Sciences* 103, no. 2 (spring 1974).

Feder, Kenneth and Michael Alan Park. *Human Antiquity: An Introduction to Physical Anthropology and Archaeology.* Mountain View, California: Mayfield Publishing Company, 1989.

Goldberg, David Theo. *Anatomy of Racism.* Minneapolis: University of Minnesota Press, 1990.

Gould, Stephen. *The Mismeasure of Man.* New York: W. W. Norton, 1981.

Greene, John. C. *Evolution and its Impact on Western Thought: The Death of Adam.* New York: The New American Library of World Literature, 1961.

Hannaford, Ivan. Race: *The History of an Idea in the West.* Washington D. C.: The Woodrow Wilson Center Press, 1996.

Harding, Sandra. *"Racial" Economy of Science.* Bloomington, Indiana: Indiana University Press, 1993.

Jacquard, Albert. *In Praise of Difference. Genetics and Human Affairs.* New York: Columbia University Press, 1984.

Jordan, Winthrop. *White over Black: American Attitudes towards the Negro, 1550-1812.* Chapel Hill: University of North Carolina Press, 1968.

Klass, Morton and Hal Hellman. *The Kinds of Mankind: An Introduction to Racism.* New York: J.B. Lippincott Company, 1971.

Lewontin, Richard. *Human Diversity.* New York: Scientific American Library, 1982.

Lewontin, Richard, Steven Rose, and Leon Kamin. *Not in Our Genes.* New York: Pantheon Books, 1984.

Li, C.C. "A Tale of Thermos Bottles; Properties of a Genetic Model for Human Intelligence." In *Intelligence, Genetic, and Environmental Influence.* Robert Cranco, ed. New York: Grune, 1971.

Mason, Philip. Common Sense about Race. New York: MacMillan, 1961.

Mead, M., T. Dobzansky, Ethel Tobach, and Robert Light. *Science and the Concept of Race.* New York: Columbia University Press, 1968.

Montagu, Ashley. *Human Heredity.* New York: A Mentor Book. World Publishing Co., 1959.

Montagu, Ashley. *Race and IQ.* New York: Oxford University Press, 1975.

Montagu, Ashley. *The Concept of Race.* New York: The Free Press; Collier; MacMillan Limited, 1963.

Montagu, Ashley. *Man's Most Dangerous Myth. The Fallacy of Race.* 4th Edition. New York: World Publishing, 1964.

Müller-Hill Benno. *Murderous Science: Elimination by scientific selection of Jews, Gypsies, and others, Germany 1933-1945.* Oxford: University Press, 1988.

Richardson, K. and D. Spears. *Race and Intelligence. The Fallacies behind the Race and IQ Controversy.* Baltimore: Penguin Books, 1972.

Riley, Vernon. *Pigmentation: Its Genesis and Biological Control.* New York: Appleton-Century Crofts, 1972.

Robins, Ashley. *Biological Perspectives on Human Pigmentation.* Cambridge Studies in Biological Anthropology. Cambridge: Cambridge University Press, 1991.

Ruse, Michael. *Darwinism Defended: A Guide to the Evolution Controversies.* London: Addison-Wesley Publishing Co., 1982.

Scheinfred, Amram. *Your Heredity and Environment.* Philadelphia: Lippincott, 1965.

Senna, Carl. *The Fallacy of I. Q.* New York: The Third Press; Joseph Opaku Publishing Co., 1973.

Shipman Pat. *The Evolution of Racism: Human Differences and the Use and Abuse of Science.* New York: Simon and Schuster, 1994.

Smedley, Audrey. *Race in North America. Origin and Evolution of a Worldview.* Boulder, Colorado: Westview Press, 1993.

Spencer, Frank. *A History of American Anthropology, 1930-1980.* New York: Academic Press, 1982.

Tobach, Ethel and Betty Rosoff. *Challenging Racism and Sexism: Alternatives to Genetic Explanations.* New York: The Feminist Press, 1994.

UNESCO. *Race and Science. The Race Question in Science.* New York: Columbia University Press, 1950.

Volpe, Peter. *Understanding Evolution.* Dubuque, Iowa: William C. Brown, 1977.

Williams, B.J. *Evolution and Human Origins: An Introduction to Physical Anthropology.* Second Edition. New York: Harper and Row, Publishers, 1979.

INDEX

137461

DEPAUL UNIVERSITY LIBRARY

3 0511 00634 5367

DATE DUE

NOV 2 4 1998		
NOV 1 0 1998		
DEC 1 6 1999		
NOV 2 3 1999		
SEP 0 5 2000		
AUG 2 8 2000		
SEP 1 5 2001		
12-14-02		
7-18-03		
SEP 2 2 2010		

Demco, Inc. 38-293